AF327927

Editors

Prof. Bruno Siciliano
Dipartimento di Ingegneria Elettrica
e Tecnologie dell'Informazione
Università degli Studi di Napoli
Federico II
Via Claudio 21, 80125 Napoli
Italy
E-mail: siciliano@unina.it

Prof. Oussama Khatib
Artificial Intelligence Laboratory
Department of Computer Science
Stanford University
Stanford, CA 94305-9010
USA
E-mail: khatib@cs.stanford.edu

For further volumes:
http://www.springer.com/series/5208

Editorial Advisory Board

Oliver Brock, TU Berlin, Germany
Herman Bruyninckx, KU Leuven, Belgium
Raja Chatila, ISIR - UPMC & CNRS, France
Henrik Christensen, Georgia Tech, USA
Peter Corke, Queensland Univ. Technology, Australia
Paolo Dario, Scuola S. Anna Pisa, Italy
Rüdiger Dillmann, Univ. Karlsruhe, Germany
Ken Goldberg, UC Berkeley, USA
John Hollerbach, Univ. Utah, USA
Makoto Kaneko, Osaka Univ., Japan
Lydia Kavraki, Rice Univ., USA
Vijay Kumar, Univ. Pennsylvania, USA
Sukhan Lee, Sungkyunkwan Univ., Korea
Frank Park, Seoul National Univ., Korea
Tim Salcudean, Univ. British Columbia, Canada
Roland Siegwart, ETH Zurich, Switzerland
Gaurav Sukhatme, Univ. Southern California, USA
Sebastian Thrun, Stanford Univ., USA
Yangsheng Xu, Chinese Univ. Hong Kong, PRC
Shin'ichi Yuta, Tsukuba Univ., Japan

STAR (Springer Tracts in Advanced Robotics) has been promoted under the auspices of EURON (European Robotics Research Network)

Jens Kober · Jan Peters

Learning Motor Skills

From Algorithms to Robot Experiments

 Springer

Dr. Jens Kober
CoR-Lab
Universität Bielefeld
Bielefeld
Germany

Honda Research Center (HRI-EU)
Offenbach/Main
Germany

jkober@cor-lab.uni-bielefeld.de

Dr. Jan Peters
Technische Universität Darmstadt
Fachgebiet Intelligente Autonome Systeme
Darmstadt
Germany

Max-Planck Institute for Intelligent Systems
Department for Empirical Inference
Tübingen
Germany

mail@jan-peters.net

ISSN 1610-7438   ISSN 1610-742X   (electronic)
ISBN 978-3-319-03193-4   ISBN 978-3-319-03194-1   (eBook)
DOI 10.1007/978-3-319-03194-1
Springer Cham Heidelberg New York Dordrecht London

Library of Congress Control Number: 2013952948

© Springer International Publishing Switzerland 2014
This work is subject to copyright. All rights are reserved by the Publisher, whether the whole or part of the material is concerned, specifically the rights of translation, reprinting, reuse of illustrations, recitation, broadcasting, reproduction on microfilms or in any other physical way, and transmission or information storage and retrieval, electronic adaptation, computer software, or by similar or dissimilar methodology now known or hereafter developed. Exempted from this legal reservation are brief excerpts in connection with reviews or scholarly analysis or material supplied specifically for the purpose of being entered and executed on a computer system, for exclusive use by the purchaser of the work. Duplication of this publication or parts thereof is permitted only under the provisions of the Copyright Law of the Publisher's location, in its current version, and permission for use must always be obtained from Springer. Permissions for use may be obtained through RightsLink at the Copyright Clearance Center. Violations are liable to prosecution under the respective Copyright Law.
The use of general descriptive names, registered names, trademarks, service marks, etc. in this publication does not imply, even in the absence of a specific statement, that such names are exempt from the relevant protective laws and regulations and therefore free for general use.
While the advice and information in this book are believed to be true and accurate at the date of publication, neither the authors nor the editors nor the publisher can accept any legal responsibility for any errors or omissions that may be made. The publisher makes no warranty, express or implied, with respect to the material contained herein.

Printed on acid-free paper

Springer is part of Springer Science+Business Media (www.springer.com)

To Our Friends and Colleagues in Tübingen

Foreword

Robotics is undergoing a major transformation in scope and dimension. From a largely dominant industrial focus, robotics is rapidly expanding into human environments and vigorously engaged in its new challenges. Interacting with, assisting, serving, and exploring with humans, the emerging robots will increasingly touch people and their lives.

Beyond its impact on physical robots, the body of knowledge robotics has produced is revealing a much wider range of applications reaching across diverse research areas and scientific disciplines, such as: biomechanics, haptics, neurosciences, virtual simulation, animation, surgery, and sensor networks among others. In return, the challenges of the new emerging areas are proving an abundant source of stimulation and insights for the field of robotics. It is indeed at the intersection of disciplines that the most striking advances happen.

The Springer Tracts in Advanced Robotics (STAR) is devoted to bringing to the research community the latest advances in the robotics field on the basis of their significance and quality. Through a wide and timely dissemination of critical research developments in robotics, our objective with this series is to promote more exchanges and collaborations among the researchers in the community and contribute to further advancements in this rapidly growing field.

The monograph by Jens Kober and Jan Peters contributes to the state of the art in reinforcement learning applied to robotics both in terms of novel algorithms and applications. A method that learns to generalize parametrized motor plans is proposed which is obtained by imitation or reinforcement learning, by adapting a small set of global parameters, and appropriate kernel-based reinforcement learning algorithms are introduced. The applications explore highly dynamic tasks and exhibit a very efficient learning process. All proposed approaches have been extensively validated with benchmarks tasks, in simulation, and on real robots. These tasks correspond to sports and games but the presented techniques are also applicable to more mundane household tasks.

Remarkably, the monograph is based on the first author's doctoral thesis, which won the 2013 EURON Georges Giralt PhD Award. A very fine addition to STAR!

Naples, Italy   Bruno Siciliano
September 2013   STAR Editor

Preface

Ever since the word "robot" was introduced to the English language by Karel Čapek's play "Rossum's Universal Robots" in 1921, robots have been expected to become part of our daily lives. In recent years, robots such as autonomous vacuum cleaners, lawn mowers, and window cleaners, as well as a huge number of toys have been made commercially available. However, a lot of additional research is required to turn robots into versatile household helpers and companions. One of the many challenges is that robots are still very specialized and cannot easily adapt to changing environments and requirements. Since the 1960s, scientists attempt to provide robots with more autonomy, adaptability, and intelligence. Research in this field is still very active but has shifted focus from reasoning based methods towards statistical machine learning. Both navigation (i.e., moving in unknown or changing environments) and motor control (i.e., coordinating movements to perform skilled actions) are important sub-tasks.

In this book, we will discuss approaches that allow robots to learn motor skills. We mainly consider tasks that need to take into account the dynamic behavior of the robot and its environment, where a kinematic movement plan is not sufficient. The presented tasks correspond to sports and games but the presented techniques will also be applicable to more mundane household tasks. Motor skills can often be represented by motor primitives. Such motor primitives encode elemental motions which can be generalized, sequenced, and combined to achieve more complex tasks. For example, a forehand and a backhand could be seen as two different motor primitives of playing table tennis. We show how motor primitives can be employed to learn motor skills on three different levels. First, we discuss how a single motor skill, represented by a motor primitive, can be learned using reinforcement learning. Second, we show how such learned motor primitives can be generalized to new situations. Finally, we present first steps towards using motor primitives in a hierarchical setting and how several motor primitives can be combined to achieve more complex tasks.

To date, there have been a number of successful applications of learning motor primitives employing imitation learning. However, many interesting motor learning problems are high-dimensional reinforcement learning problems which are often beyond the reach of current reinforcement learning methods. We review research on reinforcement learning applied to robotics and point out key challenges and important strategies to render reinforcement learning tractable. Based on these insights, we introduce novel learning approaches both for single and generalized motor skills.

For learning single motor skills, we study parametrized policy search methods and introduce a framework of reward-weighted imitation that allows us to derive both policy gradient methods and expectation-maximization (EM) inspired algorithms. We introduce a novel EM-inspired algorithm for policy learning that is particularly well-suited for motor primitives. We show that the proposed method out-performs several well-known parametrized policy search methods on an empirical benchmark both in simulation and on a real robot. We apply it in the context of motor learning and show that it can learn a complex ball-in-a-cup task on a real Barrett WAM.

In order to avoid re-learning the complete movement, such single motor skills need to be generalized to new situations. In this book, we propose a method that learns to generalize parametrized motor plans, obtained by imitation or reinforcement learning, by adapting a small set of global parameters. We employ reinforcement learning to learn the required parameters to deal with the current situation. Therefore, we introduce an appropriate kernel-based reinforcement learning algorithm. To show its feasibility, we evaluate this algorithm on a toy example and compare it to several previous approaches. Subsequently, we apply the approach to two robot tasks, i.e., the generalization of throwing movements in darts and of hitting movements in table tennis on several different real robots, i.e., a Barrett WAM, the JST-ICORP/SARCOS CBi and a Kuka KR 6.

We present first steps towards learning motor skills jointly with a higher level strategy and evaluate the approach with a target throwing task on a BioRob. Finally, we explore how several motor primitives, representing subtasks, can be combined and prioritized to achieve a more complex task. This learning framework is validated with a ball-bouncing task on a Barrett WAM.

This book contributes to the state of the art in reinforcement learning applied to robotics both in terms of novel algorithms and applications. We have introduced the *Policy learning by Weighting Exploration with the Returns* algorithm for learning single motor skills and the *Cost-regularized Kernel Regression* to generalize motor skills to new situations. The applications explore highly dynamic tasks and exhibit a very efficient learning process. All proposed approaches have been extensively validated with benchmarks tasks, in simulation, and on real robots.

Offenbach & Darmstadt   Jens Kober
September 2013   Jan Peters

Contents

Abbreviations

In this book we use the following mathematical notation throughout:

Notation	Description
$\{x_1, x_2, \ldots, x_n\}$	set with elements $x_1, x_2, \ldots, x_n$
$\mathbb{R}$	real numbers
$\mathbf{x} = [x_1, x_2, \ldots, x_n]$	a vector
x_i	the ith component of the vector $\mathbf{x}$
$\mathbf{x}^{\text{T}}$	transpose of vector
$\mathbf{A} = [\mathbf{a}_1, \mathbf{a}_2, \ldots, \mathbf{a}_n]$	a matrix
$\mathbf{a}_i$	the ith vector of the matrix $\mathbf{A}$
a_{ij}	the i, j^{th} component of the matrix $\mathbf{A}$
$\mathbf{A}^{\text{T}}$	transpose of matrix
$\mathbf{A}^{-1}$	matrix inverse
$\mathbf{A}^{+}$	matrix pseudo-inverse
$\mathbf{A}^{\frac{1}{2}}$	matrix root
$\nabla_{\theta_i} f$	derivative with respect to parameter θ_i
$\nabla_{\boldsymbol{\theta}} f$	derivative with respect to parameters θ_i
$\frac{\partial f}{\partial q}$	partial derivative
$p(\mathbf{x})$	probability density of $\mathbf{x}$
$E\{\mathbf{x}\}$	expectation of $\mathbf{x}$
$\bar{\mathbf{x}} = \langle \mathbf{x} \rangle$	sample average of $\mathbf{x}$

The following symbols are used in several sections:

Symbol	Description
t	time
$\mathbf{x}, \dot{\mathbf{x}}, \ddot{\mathbf{x}}$	task space position, velocity, acceleration
$\mathbf{q}, \dot{\mathbf{q}}, \ddot{\mathbf{q}}$	joint space position, velocity, acceleration
$\mathbf{s}_t$	state (at time t)
$\mathbf{a}_t$	action (at time t)
$\mathbf{s}_{1:T+1}$	series of states $\mathbf{s}_t$ with $t \in \{1, 2, \ldots, T+1\}$
$\mathbf{a}_{1:T}$	series of actions $\mathbf{a}_t$ with $t \in \{1, 2, \ldots, T\}$
$\boldsymbol{\tau} = [\mathbf{s}_{1:T+1}, \mathbf{a}_{1:T}]$	rollout, episode, trial
T	rollout length
$\pi\left(\mathbf{a}_t \vert \mathbf{s}_t, t\right)$	policy
$\mathbb{T}$	set of all possible paths
r_t	reward (at time t)
$\mathbf{r}_{1:T}$	series of rewards r_t with $t \in \{1, 2, \ldots, T\}$
$R\left(\tau\right)$	return
$J\left(\boldsymbol{\theta}\right)$	expected return
$D\left(p \Vert q\right)$	Kullback-Leibler divergence
$Q^{\pi}\left(\mathbf{s}, \mathbf{a}, t\right)$	value function of policy π
$\mathbf{F}\left(\boldsymbol{\theta}\right)$	Fisher information matrix
ε	exploration
H	number of rollouts
n	index of parameter
k	index of iteration
$\mathbf{g} = \nabla_{\boldsymbol{\theta}} f\left(\boldsymbol{\theta}\right)$	gradient
α	update rate
$\boldsymbol{\psi}\left(\cdot\right)$	weights
$\mathbf{c}$	centers
$\mathbf{h}$	widths
Δt	time step
λ	ridge factor
$k\left(\mathbf{s}, \mathbf{s}\right), \mathbf{K}$	kernel, Gram matrix
$\mathbf{u}$	motor command

1

Introduction

Summary. Since Issac Asimov started to write short stories about robots in the 1940s, the idea of robots as household helpers, companions and soldiers has shaped the popular view of robotics. Science fiction movies depict robots both as friends and enemies of the human race, but in both cases their capabilities far exceed the capabilities of current real robots. Simon (1965), one of the artificial intelligence (AI) research pioneers, claimed that "machines will be capable, within twenty years, of doing any work a man can do." This over-optimistic promise led to more conservative predictions. Nowadays robots slowly start to penetrate our daily lives in the form of toys and household helpers, like autonomous vacuum cleaners, lawn mowers, and window cleaners. Most other robotic helpers are still confined to research labs and industrial settings. Many tasks of our daily lives can only be performed very slowly by a robot which often has very limited generalization capabilities. Hence, all these systems are still disconnected from the expectation raised by literature and movies as well as from the dreams of AI researchers.

Especially in Japan, the need of robotic household companions has been recognized due to the aging population. One of the main challenges remains the need to adapt to changing environments in a co-inhabited household (e.g., furniture being moved, changing lighting conditions) and the need to adapt to individual requirements and expectations of the human owner. Most current products either feature a "one size fits all" approach that often is not optimal (e.g., vacuum robots that are not aware of their surrounding but use an approach for obstacle treatment, that guarantees coverage of the whole floor (BotJunkie, 2012)) or an approach that requires a setup step either in software (e.g. providing a floor map) or in hardware (e.g., by placing beacons). As an alternative one could imagine a self-learning system. In this book, we will not treat navigation problems but rather focus on learning motor skills (Wulf, 2007). We are mainly interested in motor skills that need to take into account the dynamics of the robot and its environment. For these motor skills,

J. Kober and J. Peters, *Learning Motor Skills*,
Springer Tracts in Advanced Robotics 97,
DOI: 10.1007/978-3-319-03194-1_1, © Springer International Publishing Switzerland 2014

a kinematic plan of the movement will not be sufficient to perform the task successfully. A motor skill can often be represented by a motor primitive, i.e., a representation of a single movement that is adapted to varying situations (e.g., a forehand or a backhand in table tennis that is adapted to the ball position and velocity). We focus on learning how to perform a motor primitive optimally and how to generalize it to new situations. The presented tasks correspond to games and sports, which are activities that a user might enjoy with a robot companion, but the presented techniques could also be applied to more mundane household tasks.

1.1   Motivation

Machine-learning research has resulted in a huge number of algorithms. Unfortunately most standard approaches are not directly applicable to robotics, mainly due to the inherent high dimensionality. Hence there is a need to develop approaches specifically tailored for the needs of robot learning. In this book we focus on reinforcement learning of motor primitives.

Research in reinforcement learning started in the 1950s (e.g., (Minsky, 1954; Farley and Clark, 1954; Bellman, 1957)) but has been mainly focused on theoretical problems. The algorithms are often evaluated on synthetic benchmark problems involving discrete states and actions. Probably the best known real-world applications of reinforcement learning are games, like backgammon (Tesauro, 1994) or Go (Chan et al, 1996), but also robotic applications can be found from the 1990s on (e.g., (Mahadevan and Connell, 1992; Gullapalli et al, 1994)). In contrast to many problems studied in the reinforcement learning literature, robotic problems have inherently continuous states and actions. Furthermore, experiments in robotics often deal with expensive and potentially fragile hardware and often require human supervision and intervention. These differences result in the need of adapting existing reinforcement learning approaches or developing tailored ones.

Policy search, also known as policy learning, has become an accepted alternative of value function-based reinforcement learning (Strens and Moore, 2001; Kwee et al, 2001; Peshkin, 2001; Bagnell et al, 2003; El-Fakdi et al, 2006; Taylor et al, 2007). Especially for learning motor skills in robotics, searching directly for the policy instead of solving the dual problem (i.e., finding the value function) has been shown to be promising, see Section 2.2 for a more detailed discussion. Additionally, incorporating prior knowledge in the form of the policy structure, an initial policy, or a model of the system can drastically reduce the learning time. In this book we will focus on policy search methods that employ a pre-structured policy and an initial policy.

1.2   Contributions

This book contributes to the state of the art in reinforcement learning applied to robotics both in terms of novel algorithms (Section 1.2.1) and applications (Section 1.2.2).

1.2.1   Algorithms

Reinforcement learning applied to robotics faces a number of challenges as discussed in Section 2.3. The dimensionality is inherently high and continuous. Acquiring real world samples is expensive as they require time, human supervision and potentially expensive and fragile hardware. Using models can alleviate these problems but poses different challenges. Every learning algorithm is limited by its goal specifications. Unfortunately specifying a cost or a reward function is not straightforward. Even the cost function of a simple human reaching movement is not completely understood yet (Bays and Wolpert, 2007). Inverse reinforcement learning is a promising alternative to specifying the reward function manually and has been explored for the Ball-in-a-Cup task (Section 4.3.7) in (Boularias et al, 2011).

Ideally an algorithm applied to robotics should be safe, i.e., avoid damaging the robot, and it should be fast, both in terms of convergence and computation time. Having a sample-efficient algorithm, with very few open parameters, and the ability to incorporate prior knowledge all contribute significantly to fast convergence. In this book we propose a framework of reward-weighted imitation that fits these requirements. In Section 7.1.1, we will review how the proposed algorithms meet these requirements.

Policy Learning by Weighting Exploration with the Returns (PoWER)

In Chapter 4, we introduce the Policy learning by Weighting Exploration with the Returns (PoWER) algorithm. PoWER is an expectation-maximization inspired policy search algorithm that is based on structured exploration in the parameter space. In Chapter 4, we derive a framework of reward-weighted imitation. Based on (Dayan and Hinton, 1997), we consider the return of an episode as an improper probability distribution. We maximize a lower bound of the logarithm of the expected return. Depending on the strategy of optimizing this lower bound and the exploration strategy, the framework yields several well known policy search algorithms: episodic RE-INFORCE (Williams, 1992), the policy gradient theorem (Sutton et al, 1999), episodic natural actor critic (Peters and Schaal, 2008b), as well as a generalization of the reward-weighted regression (Peters and Schaal, 2008a). Our novel algorithm, PoWER, is based on an expectation-maximization inspired optimization and a structured, state-dependent exploration. Our approach has already given rise to follow-up work in other contexts, for example,

(Vlassis et al, 2009; Kormushev et al, 2010). Theodorou et al (2010) have shown that an algorithm very similar to PoWER can also be derived from a completely different perspective, that is, the path integral approach.

Cost-regularized Kernel Regression (CrKR)

In Chapter 5, we introduce the algorithm Cost-regularized Kernel Regression (CrKR). CrKR is a non-parametric policy search approach that is particularly suited for learning meta-parameters, i.e., a limited set of parameters that influence the movement globally. In this setting, designing good parametrized policies can be challenging, a non-parametric policy offers more flexibility. We derive CrKR based on the reward-weighted regression [Peters and Schaal, 2008a, Chapter 4]. The resulting algorithm is related to Gaussian process regression and similarly yields a predictive variance, which is employed to guide the exploration in on-policy reinforcement learning. This approach is used to learn a mapping from situation to meta-parameters while the parameters of the motor primitive can still be acquired through traditional approaches.

1.2.2   Applications

In this book we show a large number of benchmark task and evaluate the approaches on robotic tasks both with simulated and real robots. The employed robots include a Barrett WAM, a BioRob, the JST-ICORP/SARCOS CBi and a Kuka KR 6. This book studies single motor skills and how these can be generalized to new situations. The presented work has contributed to the state-of-the-art of reinforcement learning applied to robotics by exploring highly dynamic tasks and exhibiting a very efficient learning process.

Single Motor Skills

Because of the curse of dimensionality, we cannot start with an arbitrary solution to learn a motor skill. Instead, we mimic the way children learn and first present an example movement for imitation learning, which is recorded using, e.g., kinesthetic teach-in. Subsequently, our reinforcement algorithm learns how to perform the skill reliably. After only realistically few episodes, the task can be regularly fulfilled and the robot shows very good average performance. We demonstrate this approach with a number of different policy representations including dynamical systems motor primitives (Ijspeert et al, 2002a) and other parametric representations. We benchmark PoWER against a variety of policy search approaches on both synthetic and robotic tasks. Finally, we demonstrate that single movement skills like the Underactuated Swing-Up and Ball-in-a-Cup can be learned on a Barrett WAM efficiently.

Generalized Motor Skills

In order to increase the efficiency of a learning process it is often advantageous to generalize a motor skill to new situations instead of re-learning it from scratch. This kind of learning often requires a non-intuitive mapping from situation to actions. To demonstrate the proposed system in a complex scenario, we have chosen the *Around the Clock* dart throwing game, table tennis, and ball throwing implemented both on simulated and real robots. In these scenarios we show that our approach performs well in a wide variety of settings, i.e. on four different real robots (namely a Barrett WAM, a BioRob, the JST-ICORP/SARCOS CBi and a Kuka KR 6), with different cost functions (both with and without secondary objectives), and with different policies in conjunction with their associated meta-parameters. The ball throwing task presents first steps towards a hierarchical reinforcement learning system.

Many robotic tasks can be decomposed into sub-tasks. However, often these sub-tasks cannot be achieved simultaneously and a dominance structure needs to be established. We evaluate initial steps towards a control law based on a set of prioritized motor primitives with a ball-bouncing task on a Barrett WAM.

1.3   Outline

The chapters in this book can largely be read independently but partially build upon results of the preceding chapters. Figure 1.1 illustrates the outline of this book. Chapter 3 describes the employed motor primitive representation that is used as policy parametrization for the evaluations throughout the book. The theoretical contributions are applicable to a wider variety of policies. Chapter 4 describes how single movement skills can be learned using policy search and employing a parametrized policy that is linear in parameters. Chapter 5 explains how single movements can be generalized to new situations using a nonparametric policy representation. The initial movement is obtained using results from Chapters 3 and 4.

Chapter 2 provides a survey on reinforcement learning applied to robotics. The survey provides an overview of techniques and applications. The particular focus lies on challenges specific to reinforcement learning in robotics and approaches to render the learning problem tractable nevertheless. This chapter is based on (Kober et al, 2013) and a preliminary version was published in (Kober and Peters, 2012).

Chapter 3 describes how the (Ijspeert et al, 2002a) representation for motor skills can be generalized for hitting and batting movements. This chapter is based on (Kober et al, 2010a).

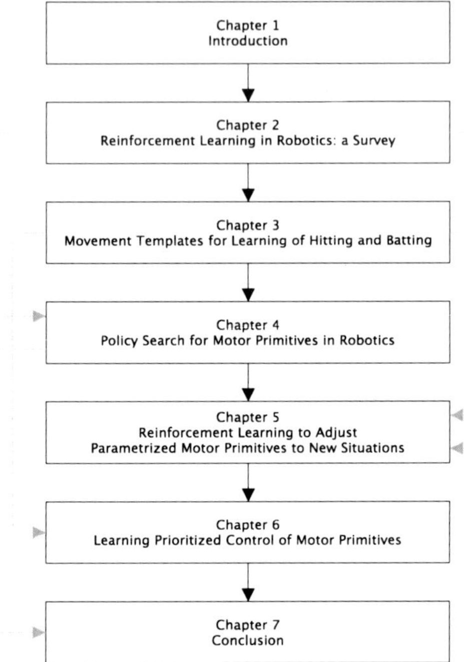

Fig. 1.1 This figure illustrates the structure of this book. Chapter 2 reviews the field of reinforcement learning applied to robotics, Chapter 3 describes the employed motor skill representation. Chapter 4 provides a framework for policy search approaches. Chapter 5 describes how movements can be generalized to new situations. Chapter 6 outlines how several motor skills can be performed simultaneously to achieve a more complex task. Chapter 1 gives an introduction and Chapter 7 provides a summary and an outlook on future work.

Chapter 4 discusses a framework of reward-weighted imitation that allows us to re-derive several well-know policy gradient approaches and to derive novel EM-inspired policy search approaches. The resulting algorithm, PoWER is evaluated in a number of benchmark and robotic tasks. This chapter is based on (Kober and Peters, 2011b) and a preliminary version of some of the work in this chapter was shown in (Kober et al, 2008; Kober and Peters, 2008, 2009, 2010).

Chapter 5 discusses how behaviors acquired by imitation and reinforcement learning as described in Chapters 3 and 4 can be generalized to novel situations using the Cost-regularized Kernel Regression. We also discuss first steps towards a hierarchical framework. The approach is evaluated on several robotic platforms. This chapter is based on (Kober et al, 2012) and a preliminary version of some of the work in this chapter was shown in (Kober et al, 2010b, 2011; Kober and Peters, 2011a).

Chapter 6 discusses first steps towards performing sub-tasks simultaneously to achieve a task. We discuss a novel learning approach that allows to learn a prioritized control law built on a set of sub-tasks represented by motor primitives. The approach is evaluated with a ball bouncing task on a Barrett WAM. This chapter is based on (Kober and Peters, submitted).

Chapter 7 concludes the book with a summary and gives an overview of open problems and an outlook on future work.

Reinforcement Learning in Robotics: A Survey

with J. Andrew Bagnell[*]

Summary. Reinforcement learning offers to robotics a framework and set of tools for the design of sophisticated and hard-to-engineer behaviors. Conversely, the challenges of robotic problems provide both inspiration, impact, and validation for developments in reinforcement learning. The relationship between disciplines has sufficient promise to be likened to that between physics and mathematics. In this article, we attempt to strengthen the links between the two research communities by providing a survey of work in reinforcement learning for behavior generation in robots. We highlight both key challenges in robot reinforcement learning as well as notable successes. We discuss how contributions tamed the complexity of the domain and study the role of algorithms, representations, and prior knowledge in achieving these successes. As a result, a particular focus of our chapter lies on the choice between model-based and model-free as well as between value function-based and policy search methods. By analyzing a simple problem in some detail we demonstrate how reinforcement learning approaches may be profitably applied, and we note throughout open questions and the tremendous potential for future research.

2.1 Introduction

A remarkable variety of problems in robotics may be naturally phrased as ones of *reinforcement learning*. Reinforcement learning (RL) enables a robot to autonomously discover an optimal behavior through trial-and-error interactions with its environment. Instead of explicitly detailing the solution to a problem, in reinforcement learning the designer of a control task provides

---

[*] Carnegie Mellon University, Robotics Institute, 5000 Forbes Avenue, Pittsburgh, PA 15213, USA.

J. Kober and J. Peters, *Learning Motor Skills,*
Springer Tracts in Advanced Robotics 97,
DOI: 10.1007/978-3-319-03194-1_2, © Springer International Publishing Switzerland 2014

feedback in terms of a scalar objective function that measures the one-step performance of the robot. Figure 2.1 illustrates the diverse set of robots that have learned tasks using reinforcement learning.

Consider, for example, attempting to train a robot to return a table tennis ball over the net (Muelling et al, 2012). In this case, the robot might make an observations of dynamic variables specifying ball position and velocity and the internal dynamics of the joint position and velocity. This might in fact capture well the *state s* of the system – providing a complete statistic for predicting future observations. The *actions a* available to the robot might be the torque sent to motors or the desired accelerations sent to an inverse dynamics control system. A function π that generates the motor commands (i.e., the actions) based on the incoming ball and current internal arm observations (i.e., the state) would be called the *policy*. A reinforcement learning problem is to find a policy that optimizes the long term sum of *rewards $R(s, a)$*; a reinforcement learning algorithm is one designed to find such a (near)-optimal policy. The reward function in this example could be based on the success of the hits as well as secondary criteria like energy consumption.

2.1.1   Reinforcement Learning in the Context of Machine Learning

In the problem of reinforcement learning, an agent explores the space of possible strategies and receives feedback on the outcome of the choices made. From this information, a "good" – or ideally optimal – policy (i.e., strategy or controller) must be deduced.

Reinforcement learning may be understood by contrasting the problem with other areas of study in machine learning. In supervised learning (Langford and Zadrozny, 2005), an agent is directly presented a sequence of independent examples of correct predictions to make in different circumstances. In imitation learning, an agent is provided demonstrations of actions of a good strategy to follow in given situations (Argall et al, 2009; Schaal, 1999).

To aid in understanding the RL problem and its relation with techniques widely used within robotics, Figure 2.2 provides a schematic illustration of two axes of problem variability: the complexity of sequential interaction and the complexity of reward structure. This hierarchy of problems, and the relations between them, is a complex one, varying in manifold attributes and difficult to condense to something like a simple linear ordering on problems. Much recent work in the machine learning community has focused on understanding the diversity and the inter-relations between problem classes. The figure should be understood in this light as providing a crude picture of the relationship between areas of machine learning research important for robotics.

Each problem subsumes those that are both below and to the left in the sense that one may always frame the simpler problem in terms of the more

(a) OBELIX robot

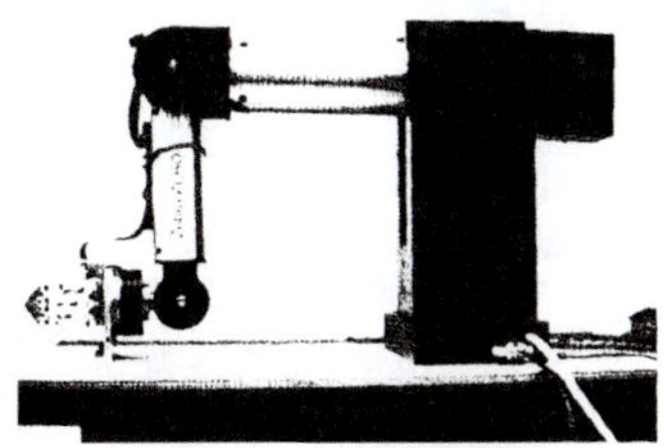

(b) Zebra Zero robot

(c) Autonomous helicopter

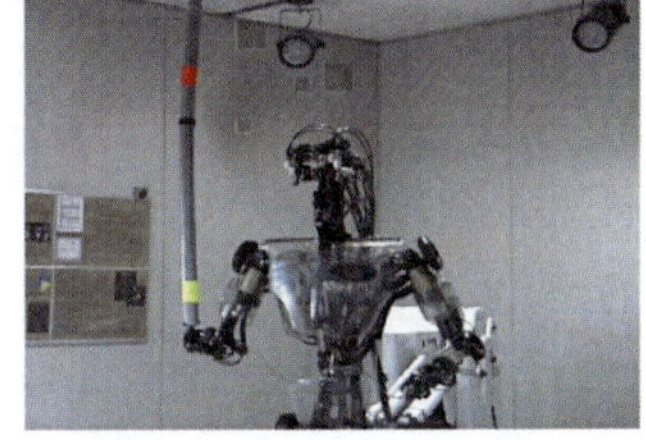

(d) Sarcos humanoid DB

Fig. 2.1 This figure illustrates a small sample of robots with behaviors that were reinforcement learned. These cover the whole range of aerial vehicles, robotic arms, autonomous vehicles, and humanoid robots. (a) The OBELIX robot is a wheeled mobile robot that learned to push boxes (Mahadevan and Connell, 1992) with a value function-based approach (Picture reprint with permission of Sridhar Mahadevan). (b) A Zebra Zero robot arm learned a peg-in-hole insertion task (Gullapalli et al, 1994) with a model-free policy gradient approach (Picture reprint with permission of Rod Grupen). (c) Carnegie Mellon's autonomous helicopter leveraged a model-based policy search approach to learn a robust flight controller (Bagnell and Schneider, 2001). (d) The Sarcos humanoid DB learned a pole-balancing task (Schaal, 1996) using forward models (Picture reprint with permission of Stefan Schaal).

complex one; note that some problems are not linearly ordered. In this sense, reinforcement learning subsumes much of the scope of classical machine learning as well as contextual bandit and imitation learning problems. Reduction algorithms (Langford and Zadrozny, 2005) are used to convert effective solutions for one class of problems into effective solutions for others, and have proven to be a key technique in machine learning.

At lower left, we find the paradigmatic problem of supervised learning, which plays a crucial role in applications as diverse as face detection and spam filtering. In these problems (including binary classification and regression), a learner's goal is to map observations (typically known as features or covariates) to actions which are usually a discrete set of classes or a real value. These problems possess no interactive component: the design and analysis of algorithms to address these problems rely on training and testing instances

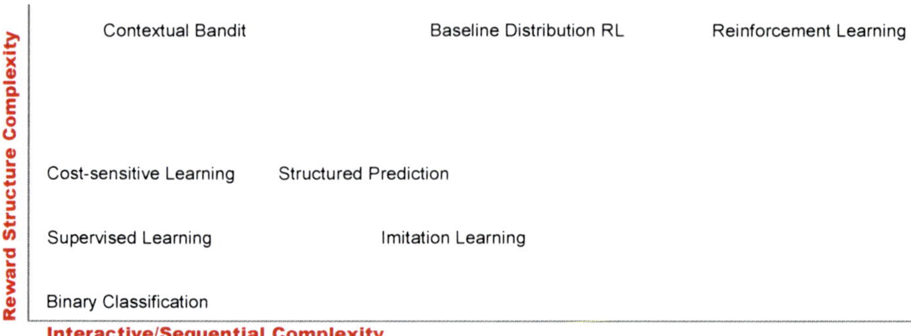

Fig. 2.2 An illustration of the inter-relations between well-studied learning problems in the literature along axes that attempt to capture both the information and complexity available in reward signals and the complexity of sequential interaction between learner and environment. Each problem subsumes those to the left and below; reduction techniques provide methods whereby harder problems (above and right) may be addressed using repeated application of algorithms built for simpler problems. (Langford and Zadrozny, 2005)

as independent and identical distributed random variables. This rules out any notion that a decision made by the learner will impact future observations: supervised learning algorithms are built to operate in a world in which every decision has no effect on the future examples considered. Further, within supervised learning scenarios, during a training phase the "correct" or preferred answer is provided to the learner, so there is no ambiguity about action choices.

More complex reward structures are also often studied: one such is known as cost-sensitive learning, where each training example and each action or prediction is annotated with a cost for making such a prediction. Learning techniques exist that reduce such problems to the simpler classification problem, and active research directly addresses such problems as they are crucial in practical learning applications.

Contextual bandit or associative reinforcement learning problems begin to address the fundamental problem of exploration-vs-exploitation, as information is provided only about a chosen action and not what-might-have-been. These find wide-spread application in problems as diverse as pharmaceutical drug discovery to ad placement on the web, and are one of the most active research areas in the field.

Problems of imitation learning and structured prediction may be seen to vary from supervised learning on the alternate dimension of sequential interaction. Structured prediction, a key technique used within computer vision and robotics, where many predictions are made in concert by leveraging inter-relations between them, may be seen as a simplified variant of imitation learning (Daumé III et al, 2009; Ross et al, 2011a). In imitation learning, we assume that an expert (for example, a human pilot) that we wish to mimic provides demonstrations of a task. While "correct answers" are provided to the learner, complexity arises because any mistake by the learner modifies the future observations from what would have been seen had the expert chosen the controls. Such problems provably lead to compounding errors and violate the basic assumption of independent examples required for successful supervised learning. In fact, in sharp contrast with supervised learning problems where only a single data-set needs to be collected, repeated interaction between learner and teacher appears to both necessary and sufficient (Ross et al, 2011b) to provide performance guarantees in both theory and practice in imitation learning problems.

Reinforcement learning embraces the full complexity of these problems by requiring both interactive, sequential prediction as in imitation learning as well as complex reward structures with only "bandit" style feedback on the actions actually chosen. It is this combination that enables so many problems of relevance to robotics to be framed in these terms; it is this same combination that makes the problem both information-theoretically and computationally hard.

We note here briefly the problem termed "Baseline Distribution RL": this is the standard RL problem with the additional benefit for the learner that it may draw initial states from a distribution provided by an expert instead of simply an initial state chosen by the problem. As we describe further in Section 2.5.1, this additional information of which states matter dramatically affects the complexity of learning.

2.1.2 *Reinforcement Learning in the Context of Optimal Control*

Reinforcement Learning (RL) is very closely related to the theory of classical optimal control, as well as dynamic programming, stochastic programming, simulation-optimization, stochastic search, and optimal stopping (Powell, 2012). Both RL and optimal control address the problem of finding an optimal policy (often also called the controller or control policy) that optimizes an objective function (i.e., the accumulated cost or reward), and both rely on the notion of a system being described by an underlying set of states, controls and a plant or model that describes transitions between states. However, optimal control assumes perfect knowledge of the system's description in the form of a model (i.e., a function T that describes what the next state of the robot will be given the current state and action). For such models, optimal control ensures strong guarantees which, nevertheless, often break down due to model and computational approximations. In contrast, reinforcement learning operates directly on measured data and rewards from interaction with the environment. Reinforcement learning research has placed great focus on addressing cases which are analytically intractable using approximations and data-driven techniques. One of the most important approaches to reinforcement learning within robotics centers on the use of classical optimal control techniques (e.g. Linear-Quadratic Regulation and Differential Dynamic Programming) to system models learned via repeated interaction with the environment (Atkeson, 1998; Bagnell and Schneider, 2001; Coates et al, 2009). A concise discussion of viewing reinforcement learning as "adaptive optimal control" is presented in (Sutton et al, 1991).

2.1.3   Reinforcement Learning in the Context of Robotics

Robotics as a reinforcement learning domain differs considerably from most well-studied reinforcement learning benchmark problems. In this article, we highlight the challenges faced in tackling these problems. Problems in robotics are often best represented with high-dimensional, continuous states and actions (note that the 10-30 dimensional continuous actions common in robot reinforcement learning are considered large (Powell, 2012)). In robotics, it is often unrealistic to assume that the true state is completely observable and noise-free. The learning system will not be able to know precisely in which state it is and even vastly different states might look very similar. Thus, robotics reinforcement learning are often modeled as partially observed, a point we take up in detail in our formal model description below. The learning system must hence use filters to estimate the true state. It is often essential to maintain the information state of the environment that not only contains the raw observations but also a notion of uncertainty on its estimates (e.g., both the mean and the variance of a Kalman filter tracking the ball in the robot table tennis example).

Experience on a real physical system is tedious to obtain, expensive and often hard to reproduce. Even getting to the same initial state is impossible for the robot table tennis system. Every single trial run, also called a roll-out, is costly and, as a result, such applications force us to focus on difficulties that do not arise as frequently in classical reinforcement learning benchmark examples. In order to learn within a reasonable time frame, suitable approximations of state, policy, value function, and/or system dynamics need to be introduced. However, while real-world experience is costly, it usually cannot be replaced by learning in simulations alone. In analytical or learned models of the system even small modeling errors can accumulate to a substantially different behavior, at least for highly dynamic tasks. Hence, algorithms need to be robust with respect to models that do not capture all the details of the real system, also referred to as *under-modeling*, and to model uncertainty. Another challenge commonly faced in robot reinforcement learning is the generation of appropriate reward functions. Rewards that guide the learning system quickly to success are needed to cope with the cost of real-world experience. This problem is called *reward shaping* (Laud, 2004) and represents a substantial manual contribution. Specifying good reward functions in robotics requires a fair amount of domain knowledge and may often be hard in practice.

Not every reinforcement learning method is equally suitable for the robotics domain. In fact, many of the methods thus far demonstrated on difficult problems have been model-based (Atkeson et al, 1997; Abbeel et al, 2007; Deisenroth and Rasmussen, 2011) and robot learning systems often employ policy search methods rather than value function-based approaches (Gullapalli et al, 1994; Miyamoto et al, 1996; Bagnell and Schneider, 2001; Kohl and Stone, 2004; Tedrake et al, 2005; Peters and Schaal, 2008a,b; Kober and Peters, 2008; Deisenroth et al, 2011). Such design choices stand in contrast to possibly the bulk of the early research in the machine learning community (Kaelbling et al, 1996; Sutton and Barto, 1998). We attempt to give a fairly complete overview on real robot reinforcement learning citing most original papers while grouping them based on the key insights employed to make the Robot Reinforcement Learning problem tractable. We isolate key insights such as choosing an appropriate representation for a value function or policy, incorporating prior knowledge, and transfer knowledge from simulations.

This chapter surveys a wide variety of tasks where reinforcement learning has been successfully applied to robotics. If a task can be phrased as an optimization problem and exhibits temporal structure, reinforcement learning can often be profitably applied to both phrase and solve that problem. The goal of this chapter is twofold. On the one hand, we hope that this chapter can provide indications for the robotics community which type of problems can be tackled by reinforcement learning and provide pointers to approaches that are promising. On the other hand, for the reinforcement learning community, this chapter can point out novel real-world test beds and remarkable opportunities for research on open questions. We focus mainly on results that were obtained on physical robots with tasks going beyond typical reinforcement learning benchmarks.

We concisely present reinforcement learning techniques in the context of robotics in Section 2.2. The challenges in applying reinforcement learning in robotics are discussed in Section 2.3. Different approaches to making reinforcement learning tractable are treated in Sections 2.4 to 2.6. In Section 2.7, the example of ball-in-a-cup is employed to highlight which of the various approaches discussed in the chapter have been particularly helpful to make such a complex task tractable. Finally, in Section 2.8, we summarize the specific problems and benefits of reinforcement learning in robotics and provide concluding thoughts on the problems and promise of reinforcement learning in robotics.

2.2   A Concise Introduction to Reinforcement Learning

In reinforcement learning, an agent tries to maximize the accumulated reward over its life-time. In an *episodic* setting, where the task is restarted after each end of an episode, the objective is to maximize the total reward per episode. If the task is on-going without a clear beginning and end, either the average reward over the whole life-time or a discounted return (i.e., a weighted average where distant rewards have less influence) can be optimized. In such reinforcement learning problems, the agent and its environment may be modeled being in a state $s \in S$ and can perform actions $a \in A$, each of which may be members of either discrete or continuous sets and can be multi-dimensional. A state s contains all relevant information about the current situation to predict future states (or observables); an example would be the current position of a robot in a navigation task[1]. An action a is used to control (or change) the state of the system. For example, in the navigation task we could have the actions corresponding to torques applied to the wheels. For every step, the agent also gets a reward R, which is a scalar value and assumed to be a function of the state and observation. (It may equally be modeled as a random variable that depends on only these variables.) In the navigation task, a possible reward could be designed based on the energy costs for taken actions and rewards for reaching targets. The goal of reinforcement learning is to find a mapping from states to actions, called policy π, that picks actions a in given states s maximizing the cumulative expected reward. The policy π is either deterministic or probabilistic. The former always uses the exact same action for a given state in the form $a = \pi(s)$, the later draws a sample from a distribution over actions when it encounters a state, i.e., $a \sim \pi(s, a) = P(a|s)$. The reinforcement learning agent needs to discover the relations between states, actions, and rewards. Hence exploration is required which can either be directly embedded in the policy or performed separately and only as part of the learning process.

Classical reinforcement learning approaches are based on the assumption that we have a *Markov Decision Process* (MDP) consisting of the set of states S, set of actions A, the rewards R and transition probabilities T that capture the dynamics of a system. Transition probabilities (or densities in the continuous state case) $T(s', a, s) = P(s'|s, a)$ describe the effects of the actions on the state. Transition probabilities generalize the notion of deterministic dynamics to allow for modeling outcomes are uncertain even given full state. The Markov property requires that the next state s' and the reward only depend on the previous state s and action a (Sutton and Barto, 1998), and not on additional information about the past states or actions. In a sense, the Markov property recapitulates the idea of state – a state is a sufficient statistic for predicting the future, rendering previous observations

---

[1] When only observations but not the complete state is available, the sufficient statistics of the filter can alternatively serve as state s. Such a state is often called information or belief state.

irrelevant. In general in robotics, we may only be able to find some approximate notion of state.

Different types of reward functions are commonly used, including rewards depending only on the current state $R = R(s)$, rewards depending on the current state and action $R = R(s, a)$, and rewards including the transitions $R = R(s', a, s)$. Most of the theoretical guarantees only hold if the problem adheres to a Markov structure, however in practice, many approaches work very well for many problems that do not fulfill this requirement.

2.2.1 Goals of Reinforcement Learning

The goal of reinforcement learning is to discover an optimal policy π^* that maps states (or observations) to actions so as to maximize the expected return J, which corresponds to the cumulative expected reward. There are different models of optimal behavior (Kaelbling et al, 1996) which result in different definitions of the expected return. A finite-horizon model only attempts to maximize the expected reward for the horizon H, i.e., the next H (time-) steps h

$$J = E \left\{ \sum_{h=0}^{H} R_h \right\}.$$

This setting can also be applied to model problems where it is known how many steps are remaining.

Alternatively, future rewards can be discounted by a discount factor γ (with $0 \leq \gamma < 1$)

$$J = E \left\{ \sum_{h=0}^{\infty} \gamma^h R_h \right\}.$$

This is the setting most frequently discussed in classical reinforcement learning texts. The parameter γ affects how much the future is taken into account and needs to be tuned manually. As illustrated in (Kaelbling et al, 1996), this parameter often qualitatively changes the form of the optimal solution. Policies designed by optimizing with small γ are myopic and greedy, and may lead to poor performance if we actually care about longer term rewards. It is straightforward to show that the optimal control law can be unstable if the discount factor is too low (e.g., it is not difficult to show this destabilization even for discounted linear quadratic regulation problems). Hence, discounted formulations are frequently inadmissible in robot control.

In the limit when γ approaches 1, the metric approaches what is known as the average-reward criterion (Bertsekas, 1995),

$$J = \lim_{H \to \infty} E \left\{ \frac{1}{H} \sum_{h=0}^{H} R_h \right\}.$$

This setting has the problem that it cannot distinguish between policies that initially gain a transient of large rewards and those that do not. This transient phase, also called prefix, is dominated by the rewards obtained in the long run. If a policy accomplishes both an optimal prefix as well as an optimal long-term behavior, it is called bias optimal Lewis and Puterman (2001). An example in robotics would be the transient phase during the start of a rhythmic movement, where many policies will accomplish the same long-term reward but differ substantially in the transient (e.g., there are many ways of starting the same gait in dynamic legged locomotion) allowing for room for improvement in practical application.

In real-world domains, the shortcomings of the discounted formulation are often more critical than those of the average reward setting as stable behavior is often more important than a good transient (Peters et al, 2004). We also often encounter an episodic control task, where the task runs only for H time-steps and then reset (potentially by human intervention) and started over. This horizon, H, may be arbitrarily large, as long as the expected reward over the episode can be guaranteed to converge. As such episodic tasks are probably the most frequent ones, finite-horizon models are often the most relevant.

Two natural goals arise for the learner. In the first, we attempt to find an optimal strategy at the end of a phase of training or interaction. In the second, the goal is to maximize the reward over the whole time the robot is interacting with the world.

In contrast to supervised learning, the learner must first discover its environment and is not told the optimal action it needs to take. To gain information about the rewards and the behavior of the system, the agent needs to explore by considering previously unused actions or actions it is uncertain about. It needs to decide whether to play it safe and stick to well known actions with (moderately) high rewards or to dare trying new things in order to discover new strategies with an even higher reward. This problem is commonly known as the *exploration-exploitation trade-off*.

In principle, reinforcement learning algorithms for Markov Decision Processes with performance guarantees are known (Kakade, 2003; Kearns and Singh, 2002; Brafman and Tennenholtz, 2002) with polynomial scaling in the size of the state and action spaces, an additive error term, as well as in the horizon length (or a suitable substitute including the discount factor or "mixing time" (Kearns and Singh, 2002)). However, state spaces in robotics problems are often tremendously large as they scale exponentially in the number of state variables and often are continuous. This challenge of exponential growth is often referred to as the *curse of dimensionality* (Bellman, 1957) (also discussed in Section 2.3.1).

Off-policy methods learn independent of the employed policy, i.e., an explorative strategy that is different from the desired final policy can be employed during the learning process. *On-policy* methods collect sample information about the environment using the current policy. As a result, exploration must be built into the policy and determines the speed of the policy improvements. Such exploration and the performance of the policy can result in an exploration-exploitation trade-off between long- and short-term improvement of the policy. Modeling exploration models with probability distributions has surprising implications, e.g., stochastic policies have been shown to be the optimal stationary policies for selected problems (Sutton et al, 1999; Jaakkola et al, 1993) and can even break the curse of dimensionality (Rust, 1997). Furthermore, stochastic policies often allow the derivation of new policy update steps with surprising ease.

The agent needs to determine a correlation between actions and reward signals. An action taken does not have to have an immediate effect on the reward but can also influence a reward in the distant future. The difficulty in assigning credit for rewards is directly related to the horizon or mixing time of the problem. It also increases with the dimensionality of the actions as not all parts of the action may contribute equally.

The classical reinforcement learning setup is a MDP where additionally to the states S, actions A, and rewards R we also have transition probabilities $T(s', a, s)$. Here, the reward is modeled as a reward function $R(s, a)$. If both the transition probabilities and reward function are known, this can be seen as an optimal control problem (Powell, 2012).

2.2.2   Reinforcement Learning in the Average Reward Setting

We focus on the average-reward model in this section. Similar derivations exist for the finite horizon and discounted reward cases. In many instances, the average-reward case is often more suitable in a robotic setting as we do not have to choose a discount factor and we do not have to explicitly consider time in the derivation.

To make a policy able to be optimized by continuous optimization techniques, we write a policy as a conditional probability distribution $\pi(s, a) = P(a|s)$. Below, we consider restricted policies that are paramertized by a vector θ. In reinforcement learning, the policy is usually considered to be stationary and memory-less. Reinforcement learning and optimal control aim at finding the optimal policy π^* or equivalent policy parameters θ^* which maximize the average return $J(\pi) = \sum_{s,a} \mu^\pi(s)\pi(s, a)R(s, a)$ where μ^π is the stationary state distribution generated by policy π acting in the environment, i.e., the MDP. It can be shown (Puterman, 1994) that such policies that map states (even deterministically) to actions are sufficient to ensure optimality in this setting – a policy needs neither to remember previous states visited, actions taken, or the particular time step. For simplicity and to ease exposition, we assume that this distribution is unique. Markov Decision Processes where this fails (i.e., non-ergodic processes) require more care in analysis, but similar results exist (Puterman, 1994). The transitions between states s caused by actions a are modeled as $T(s, a, s') = P(s'|s, a)$. We can then frame the control problem as an optimization of

$$\max_{\pi} J\left(\pi\right) = \sum_{s,a} \mu^{\pi}(s)\pi(s,a)R(s,a), \tag{2.1}$$

$$\text{s.t. } \mu^{\pi}(s') = \sum_{s,a} \mu^{\pi}(s)\pi(s,a)T(s,a,s'), \ \forall s' \in S, \tag{2.2}$$

$$1 = \sum_{s,a} \mu^{\pi}(s)\pi(s,a) \tag{2.3}$$

$$\pi(s,a) \geq 0, \ \forall s \in S, a \in A.$$

Here, Equation (2.2) defines stationarity of the state distributions μ^{π} (i.e., it ensures that it is well defined) and Equation (2.3) ensures a proper state-action probability distribution. This optimization problem can be tackled in two substantially different ways (Bellman, 1967, 1971). We can search the optimal solution directly in this original, primal problem or we can optimize in the Lagrange dual formulation. Optimizing in the primal formulation is known as *policy search* in reinforcement learning while searching in the dual formulation is known as a *value function-based approach.*

Value Function Approaches

Much of the reinforcement learning literature has focused on solving the optimization problem in Equations (2.1-2.3) in its dual form (Gordon, 1999; Puterman, 1994)[2]. Using Lagrange multipliers $V^{\pi}(s')$ and $\bar{R}$, we can express the Lagrangian of the problem by

$$L = \sum_{s,a} \mu^{\pi}(s)\pi(s,a)R(s,a)$$

$$+ \sum_{s'} V^{\pi}(s') \left[\sum_{s,a} \mu^{\pi}(s)\pi(s,a)T(s,a,s') - \mu^{\pi}(s') \right]$$

$$+ \bar{R} \left[1 - \sum_{s,a} \mu^{\pi}(s)\pi(s,a) \right]$$

$$= \sum_{s,a} \mu^{\pi}(s)\pi(s,a) \left[R(s,a) + \sum_{s'} V^{\pi}(s')T(s,a,s') - \bar{R} \right]$$

$$- \sum_{s'} V^{\pi}(s')\mu^{\pi}(s') \underbrace{\sum_{a'} \pi(s',a')}_{=1} + \bar{R}.$$

Using the property $\sum_{s',a'} V(s')\mu^{\pi}(s')\pi(s',a') = \sum_{s,a} V(s)\mu^{\pi}(s)\pi(s,a)$, we can obtain the Karush-Kuhn-Tucker conditions (Kuhn and Tucker, 1950) by differentiating with respect to $\mu^{\pi}(s)\pi(s,a)$ which yields extrema at

$$\partial_{\mu^{\pi}\pi}L = R(s,a) + \sum_{s'} V^{\pi}(s')T(s,a,s') - \bar{R} - V^{\pi}(s) = 0.$$

---

[2] For historical reasons, what we call the dual is often referred to in the literature as the primal. We argue that problem of optimizing expected reward is the fundamental problem, and values are an auxiliary concept.

This statement implies that there are as many equations as the number of states multiplied by the number of actions. For each state there can be one or several optimal actions a^* that result in the *same* maximal value, and, hence, can be written in terms of the optimal action a^* as $V^{\pi^*}(s) = R(s, a^*) - \bar{R} + \sum_{s'} V^{\pi^*}(s')T(s, a^*, s')$. As a^* is generated by the same optimal policy π^*, we know the condition for the multipliers at optimality is

$$V^*(s) = \max_{a^*} \left[R(s, a^*) - \bar{R} + \sum_{s'} V^*(s')T(s, a^*, s') \right], \qquad (2.4)$$

where $V^*(s)$ is a shorthand notation for $V^{\pi^*}(s)$. This statement is equivalent to the *Bellman Principle of Optimality* (Bellman, 1957)[3] that states "An optimal policy has the property that whatever the initial state and initial decision are, the remaining decisions must constitute an optimal policy with regard to the state resulting from the first decision." Thus, we have to perform an optimal action a^*, and, subsequently, follow the optimal policy π^* in order to achieve a global optimum. When evaluating Equation (2.4), we realize that optimal value function $V^*(s)$ corresponds to the long term additional reward, beyond the average reward $\bar{R}$, gained by starting in state s while taking optimal actions a^* (according to the optimal policy π^*). This principle of optimality has also been crucial in enabling the field of optimal control (Kirk, 1970).

Hence, we have a dual formulation of the original problem that serves as condition for optimality. Many traditional reinforcement learning approaches are based on identifying (possibly approximate) solutions to this equation, and are known as *value function methods*. Instead of directly learning a policy, they first approximate the Lagrangian multipliers $V^*(s)$, also called the value function, and use it to reconstruct the optimal policy. The value function $V^\pi(s)$ is defined equivalently, however instead of always taking the optimal action a^*, the action a is picked according to a policy π

$$V^\pi(s) = \sum_{a} \pi(s, a) \left(R(s, a) - \bar{R} + \sum_{s'} V^\pi(s')T(s, a, s') \right).$$

Instead of the value function $V^\pi(s)$ many algorithms rely on the state-action value function $Q^\pi(s, a)$ instead, which has advantages for determining the optimal policy as shown below. This function is defined as

$$Q^\pi(s, a) = R(s, a) - \bar{R} + \sum_{s'} V^\pi(s')T(s, a, s').$$

In contrast to the value function $V^\pi(s)$, the state-action value function $Q^\pi(s, a)$ explicitly contains the information about the effects of a particular action. The optimal state-action value function is

---

[3] This optimality principle was originally formulated for a setting with discrete time steps and continuous states and actions but is also applicable for discrete states and actions.

$$Q^*(s,a) = R(s,a) - \bar{R} + \sum_{s'} V^*(s')T(s,a,s').$$

$$= R(s,a) - \bar{R} + \sum_{s'} \left(\max_{a'} Q^*(s',a') \right) T(s,a,s').$$

It can be shown that an optimal, deterministic policy $\pi^*(s)$ can be reconstructed by always picking the action a^* in the current state that leads to the state s with the highest value $V^*(s)$

$$\pi^*(s) = \arg\max_a \left(R(s,a) - \bar{R} + \sum_{s'} V^*(s')T(s,a,s') \right)$$

If the optimal value function $V^*(s')$ and the transition probabilities $T(s,a,s')$ for the following states are known, determining the optimal policy is straightforward in a setting with discrete actions as an exhaustive search is possible. For continuous spaces, determining the optimal action a^* is an optimization problem in itself. If both states and actions are discrete, the value function and the policy may, in principle, be represented by tables and picking the appropriate action is reduced to a look-up. For large or continuous spaces representing the value function as a table becomes intractable. Function approximation is employed to find a lower dimensional representation that matches the real value function as closely as possible, as discussed in Section 2.2.4. Using the state-action value function $Q^*(s,a)$ instead of the value function $V^*(s)$

$$\pi^*(s) = \arg\max_a \left(Q^*(s,a) \right),$$

avoids having to calculate the weighted sum over the successor states, and hence no knowledge of the transition function is required.

A wide variety of methods of value function based reinforcement learning algorithms that attempt to estimate $V^*(s)$ or $Q^*(s,a)$ have been developed and can be split mainly into three classes: (i) dynamic programming-based optimal control approaches such as policy iteration or value iteration, (ii) rollout-based Monte Carlo methods and (iii) temporal difference methods such as TD(λ) (Temporal Difference learning), Q-learning, and SARSA (State-Action-Reward-State-Action).

Dynamic Programming-Based Methods

Require a model of the transition probabilities $T(s',a,s)$ and the reward function $R(s,a)$ to calculate the value function. The model does not necessarily need to be predetermined but can also be learned from data, potentially incrementally. Such methods are called *model-based*. Typical methods include policy iteration and value iteration.

Policy iteration alternates between the two phases of policy evaluation and policy improvement. The approach is initialized with an arbitrary policy. Policy evaluation determines the value function for the current policy. Each state is visited and its value is updated based on the current value estimates of its successor states, the associated transition probabilities, as well as the policy. This procedure is repeated until the value function converges to a fixed point, which corresponds to the true value function. Policy improvement greedily selects the best action in every state according to the value function as shown above. The two steps of policy evaluation and policy improvement are iterated until the policy does not change any longer.

Policy iteration only updates the policy once the policy evaluation step has converged. In contrast, value iteration combines the steps of policy evaluation and policy improvement by directly updating the value function based on Eq. (2.4) every time a state is updated.

Monte Carlo Methods

Use sampling in order to estimate the value function. This procedure can be used to replace the policy evaluation step of the dynamic programming-based methods above. Monte Carlo methods are *model-free*, i.e., they do not need an explicit transition function. They perform roll-outs by executing the current policy on the system, hence operating on-policy. The frequencies of transitions and rewards are kept track of and are used to form estimates of the value function. For example, in an episodic setting the state-action value of a given state action pair can be estimated by averaging all the returns that were received when starting from them.

Temporal Difference Methods

Unlike Monte Carlo methods, do not have to wait until an estimate of the return is available (i.e., at the end of an episode) to update the value function. Rather, they use temporal errors and only have to wait until the next time step. The temporal error is the difference between the old estimate and a new estimate of the value function, taking into account the reward received in the current sample. These updates are done iterativley and, in contrast to dynamic programming methods, only take into account the sampled successor states rather than the complete distributions over successor states. Like the Monte Carlo methods, these methods are model-free, as they do not use a model of the transition function to determine the value function. In this setting, the value function cannot be calculated analytically but has to be estimated from sampled transitions in the MDP. For example, the value function could be updated iteratively by

$$V'(s) = V(s) + \alpha \left(R(s, a) - \bar{R} + V(s') - V(s) \right),$$

where $V(s)$ is the old estimate of the value function, $V'(s)$ the updated one, and α is a learning rate. This update step is called the TD(0)-algorithm in the discounted reward case. In order to perform action selection a model of the transition function is still required.

The equivalent temporal difference learning algorithm for state-action value functions is the average reward case version of SARSA with

$$Q'(s,a) = Q(s,a) + \alpha \left(R(s,a) - \bar{R} + Q(s',a') - Q(s,a) \right),$$

where $Q(s,a)$ is the old estimate of the state-action value function and $Q'(s,a)$ the updated one. This algorithm is on-policy as both the current action a as well as the subsequent action a' are chosen according to the current policy π. The off-policy variant is called R-learning (Schwartz, 1993), which is closely related to Q-learning, with the updates

$$Q'(s,a) = Q(s,a) + \alpha \left(R(s,a) - \bar{R} + \max_{a'} Q(s',a') - Q(s,a) \right).$$

These methods do not require a model of the transition function for determining the deterministic optimal policy $\pi^*(s)$. H-learning (Tadepalli and Ok, 1994) is a related method that estimates a model of the transition probabilities and the reward function in order to perform updates that are reminiscent of value iteration.

An overview of publications using value function based methods is presented in Table 2.1. Here, model-based methods refers to all methods that employ a predetermined or a learned model of system dynamics.

Policy Search

The primal formulation of the problem in terms of policy rather then value offers many features relevant to robotics. It allows for a natural integration of expert knowledge, e.g., through both structure and initializations of the policy. It allows domain-appropriate pre-structuring of the policy in an approximate form without changing the original problem. Optimal policies often have many fewer parameters than optimal value functions. For example, in linear quadratic control, the value function has quadratically many parameters in the dimensionality of the state-variables while the policy requires only linearly many parameters. Local search in policy space can directly lead to good results as exhibited by early hill-climbing approaches (Kirk, 1970), as well as more recent successes (see Table 2.2). Additional constraints can be incorporated naturally, e.g., regularizing the change in the path distribution. As a result, policy search often appears more natural to robotics.

Nevertheless, policy search has been considered the harder problem for a long time as the optimal solution cannot directly be determined from Equations (2.1-2.3) while the solution of the dual problem leveraging *Bellman Principle of Optimality* (Bellman, 1957) enables dynamic programming based solutions.

Notwithstanding this, in robotics, policy search has recently become an important alternative to value function based methods due to better scalability as well as the convergence problems of approximate value function

Table 2.1 This table illustrates different value function based reinforcement learning methods employed for robotic tasks (both average and discounted reward cases) and associated publications.

VALUE FUNCTION APPROACHES

Approach	Employed by...
Model-Based	Bakker et al (2006); Hester et al (2010, 2012); Kalmár et al (1998); Martínez-Marín and Duckett (2005); Schaal (1996); Touzet (1997)
Model-Free	Asada et al (1996); Bakker et al (2003); Benbrahim et al (1992); Benbrahim and Franklin (1997); Birdwell and Livingston (2007); Bitzer et al (2010); Conn and Peters II (2007); Duan et al (2007, 2008); Fagg et al (1998); Gaskett et al (2000); Gräve et al (2010); Hafner and Riedmiller (2007); Huang and Weng (2002); Huber and Grupen (1997); Ilg et al (1999); Katz et al (2008); Kimura et al (2001); Kirchner (1997); Konidaris et al (2011a, 2012); Kroemer et al (2009, 2010); Kwok and Fox (2004); Latzke et al (2007); Mahadevan and Connell (1992); Matarić (1997); Morimoto and Doya (2001); Nemec et al (2009, 2010); Oßwald et al (2010); Paletta et al (2007); Pendrith (1999); Platt et al (2006); Riedmiller et al (2009); Rottmann et al (2007); Smart and Kaelbling (1998, 2002); Soni and Singh (2006); Tamošiunaitė et al (2011); Thrun (1995); Tokic et al (2009); Touzet (1997); Uchibe et al (1998); Wang et al (2006); Willgoss and Iqbal (1999)

methods (see Sections 2.2.3 and 2.4.2). Most policy search methods optimize locally around existing policies π, parametrized by a set of policy parameters θ_i, by computing changes in the policy parameters $\Delta\theta_i$ that will increase the expected return and results in iterative updates of the form

$$\theta_{i+1} = \theta_i + \Delta\theta_i.$$

The computation of the policy update is the key step here and a variety of updates have been proposed ranging from pairwise comparisons (Strens and Moore, 2001; Ng et al, 2004a) over gradient estimation using finite policy differences (Geng et al, 2006; Kohl and Stone, 2004; Mitsunaga et al, 2005; Roberts et al, 2010; Sato et al, 2002; Tedrake et al, 2005), and general stochastic optimization methods (such as Nelder-Mead (Bagnell and Schneider, 2001), cross entropy (Rubinstein and Kroese, 2004) and population-based methods (Goldberg, 1989)) to approaches coming from optimal control such as differential dynamic programming (DDP) (Atkeson, 1998) and multiple shooting approaches (Betts, 2001). We may broadly break down policy-search methods into "black box" and "white box" methods. Black box methods are general stochastic optimization algorithms (Spall, 2003) using only the expected return of policies, estimated by sampling, and do not leverage any of the internal structure of the RL problem. These may be very sophisticated techniques (Tesch et al, 2011) that use response surface estimates and bandit-like strategies to achieve good performance. White box methods take advantage of some of additional structure within the reinforcement learning domain, including, for instance, the (approximate) Markov structure of problems, developing approximate models, value-function estimates when available (Peters and Schaal, 2008c), or even simply the causal ordering of actions and rewards. A major open issue within the field is the relative merits of the these two approaches: in principle, white box methods leverage more information, but with the exception of models (which have been demonstrated repeatedly to often make tremendous performance improvements, see Section 2.6), the performance gains are traded-off with additional assumptions that may be violated and less mature optimization algorithms. Some recent work including (Stulp and Sigaud, 2012; Tesch et al, 2011) suggest that much of the benefit of policy search is achieved by black-box methods.

Some of the most popular white-box general reinforcement learning techniques that have translated particularly well into the domain of robotics include: (i) policy gradient approaches based on likelihood-ratio estimation (Sutton et al, 1999), (ii) policy updates inspired by expectation-maximization (Toussaint et al, 2010), and (iii) the path integral methods (Kappen, 2005).

Let us briefly take a closer look at gradient-based approaches first. The updates of the policy parameters are based on a hill-climbing approach, that is following the gradient of the expected return J for a defined step-size α

$$\theta_{i+1} = \theta_i + \alpha \nabla_\theta J.$$

Different methods exist for estimating the gradient $\nabla_\theta J$ and many algorithms require tuning of the step-size α.

In *finite difference gradients* P perturbed policy parameters are evaluated to obtain an estimate of the gradient. Here we have $\Delta \hat{J}_p \approx J(\theta_i + \Delta \theta_p) - J_{\mathrm{ref}}$, where $p = [1..P]$ are the individual perturbations, $\Delta \hat{J}_p$ the estimate of their influence on the return, and J_{ref} is a reference return, e.g., the return of the unperturbed parameters. The gradient can now be estimated by linear regression

$$\nabla_\theta J \approx \left(\Delta \Theta^{\mathrm{T}} \Delta \Theta \right)^{-1} \Delta \Theta^{\mathrm{T}} \Delta \hat{\mathbf{J}},$$

where the matrix $\Delta \Theta$ contains all the stacked samples of the perturbations $\Delta \theta_p$ and $\Delta \hat{\mathbf{J}}$ contains the corresponding $\Delta \hat{J}_p$. In order to estimate the gradient the number of perturbations needs to be at least as large as the number of parameters. The approach is very straightforward and even applicable to policies that are not differentiable. However, it is usually considered to be very noisy and inefficient. For the finite difference approach tuning the step-size α for the update, the number of perturbations P, and the type and magnitude of perturbations are all critical tuning factors.

Likelihood ratio methods rely on the insight that in an episodic setting where the episodes τ are generated according to the distribution

$P^\theta(\tau) = P(\tau|\theta)$ with the return of an episode $J^\tau = \sum_{h=1}^{H} R_h$ and number of steps H the expected return for a set of policy parameter θ can be expressed as

$$J^\theta = \sum_\tau P^\theta(\tau) J^\tau. \tag{2.5}$$

The gradient of the episode distribution can be written as[4]

$$\nabla_\theta P^\theta(\tau) = P^\theta(\tau) \nabla_\theta \log P^\theta(\tau), \tag{2.6}$$

which is commonly known as the likelihood ratio or *REINFORCE* (Williams, 1992) trick. Combining Equations (2.5) and (2.6) we get the gradient of the expected return in the form

$$\nabla_\theta J^\theta = \sum_\tau \nabla_\theta P^\theta(\tau) J^\tau = \sum_\tau P^\theta(\tau) \nabla_\theta \log P^\theta(\tau) J^\tau$$

$$= E\left\{ \nabla_\theta \log P^\theta(\tau) J^\tau \right\}.$$

If we have a stochastic policy $\pi^\theta(s, a)$ that generates the episodes τ, we do not need to keep track of the probabilities of the episodes but can directly express the gradient in terms of the policy as $\nabla_\theta \log P^\theta(\tau) = \sum_{h=1}^{H} \nabla_\theta \log \pi^\theta(s, a)$. Finally the gradient of the expected return with respect to the policy parameters can be estimated as

$$\nabla_\theta J^\theta = E\left\{ \left(\sum_{h=1}^{H} \nabla_\theta \log \pi^\theta(s_h, a_h) \right) J^\tau \right\}.$$

If we now take into account that rewards at the beginning of an episode cannot be caused by actions taken at the end of an episode, we can replace the return of the episode J^τ by the state-action value function $Q^\pi(s, a)$ and get (Peters and Schaal, 2008c)

$$\nabla_\theta J^\theta = E\left\{ \sum_{h=1}^{H} \nabla_\theta \log \pi^\theta(s_h, a_h) Q^\pi(s_h, a_h) \right\},$$

which is equivalent to the *policy gradient theorem* (Sutton et al, 1999). In practice, it is often advisable to subtract a reference J_{ref}, also called baseline, from the return of the episode J^τ or the state-action value function $Q^\pi(s, a)$ respectively to get better estimates, similar to the finite difference approach. In these settings, the exploration is automatically taken care of by the stochastic policy.

Initial gradient-based approaches such as finite differences gradients or REINFORCE (REward Increment = Nonnegative Factor times Offset Reinforcement times Characteristic Eligibility) (Williams, 1992) have been rather

---

[4] From multi-variate calculus we have $\nabla_\theta \log P^\theta(\tau) = \nabla_\theta P^\theta(\tau)/P^\theta(\tau)$.

slow. The weight perturbation algorithm is related to REINFORCE but can deal with non-Gaussian distributions which significantly improves the signal to noise ratio of the gradient (Roberts et al, 2010). Recent natural policy gradient approaches (Peters and Schaal, 2008c,b) have allowed for faster convergence which may be advantageous for robotics as it reduces the learning time and required real-world interactions.

A different class of safe and fast policy search methods, that are inspired by expectation-maximization, can be derived when the reward is treated as an improper probability distribution (Dayan and Hinton, 1997). Some of these approaches have proven successful in robotics, e.g., reward-weighted regression (Peters and Schaal, 2008a), Policy Learning by Weighting Exploration with the Returns (Kober and Peters, 2008), Monte Carlo Expectation-Maximization(Vlassis et al, 2009), and Cost-regularized Kernel Regression (Kober et al, 2010b). Algorithms with closely related update rules can also be derived from different perspectives including Policy Improvements with Path Integrals (Theodorou et al, 2010) and Relative Entropy Policy Search (Peters et al, 2010a).

Finally, the Policy Search by Dynamic Programming (Bagnell et al, 2003) method is a general strategy that combines policy search with the principle of optimality. The approach learns a non-stationary policy backward in time like dynamic programming methods, but does not attempt to enforce the Bellman equation and the resulting approximation instabilities (See Section 2.2.4). The resulting approach provides some of the strongest guarantees that are currently known under function approximation and limited observability It has been demonstrated in learning walking controllers and in finding near-optimal trajectories for map exploration (Kollar and Roy, 2008). The resulting method is more expensive than the value function methods because it scales quadratically in the effective time horizon of the problem. Like DDP methods (Atkeson, 1998), it is tied to a non-stationary (time-varying) policy.

An overview of publications using policy search methods is presented in Table 2.2.

One of the key open issues in the field is determining when it is appropriate to use each of these methods. Some approaches leverage significant structure specific to the RL problem (e.g. (Theodorou et al, 2010)), including reward structure, Markovanity, causality of reward signals (Williams, 1992), and value-function estimates when available (Peters and Schaal, 2008c). Others embed policy search as a generic, black-box, problem of stochastic optimization (Bagnell and Schneider, 2001; Lizotte et al, 2007; Kuindersma et al, 2011; Tesch et al, 2011). Significant open questions remain regarding which methods are best in which circumstances and further, at an even more basic level, how effective leveraging the kinds of problem structures mentioned above are in practice.

Table 2.2 This table illustrates different policy search reinforcement learning methods employed for robotic tasks and associated publications

POLICY SEARCH

Approach	Employed by...
Gradient	Deisenroth and Rasmussen (2011); Deisenroth et al (2011); Endo et al (2008); Fidelman and Stone (2004); Geng et al (2006); Guenter et al (2007); Gullapalli et al (1994); Hailu and Sommer (1998); Ko et al (2007); Kohl and Stone (2004); Kolter and Ng (2009a); Michels et al (2005); Mitsunaga et al (2005); Miyamoto et al (1996); Ng et al (2004a,b); Peters and Schaal (2008c,b); Roberts et al (2010); Rosenstein and Barto (2004); Tamei and Shibata (2009); Tedrake (2004); Tedrake et al (2005)
Other	Abbeel et al (2006, 2007); Atkeson and Schaal (1997); Atkeson (1998); Bagnell and Schneider (2001); Bagnell (2004); Buchli et al (2011); Coates et al (2009); Daniel et al (2012); Donnart and Meyer (1996); Dorigo and Colombetti (1993); Erden and Leblebicioğlu (2008); Kalakrishnan et al (2011); Kober and Peters (2008); Kober et al (2010b); Kolter et al (2008); Kuindersma et al (2011); Lizotte et al (2007); Mataric (1994); Pastor et al (2011); Peters and Schaal (2008a); Peters et al (2010a); Schaal and Atkeson (1994); Stulp et al (2011); Svinin et al (2001); Tamošiūnaitė et al (2011); Yasuda and Ohkura (2008); Youssef (2005)

2.2.3   Value Function Approaches versus Policy Search

Some methods attempt to find a value function or policy which eventually can be employed without significant further computation, whereas others (e.g., the roll-out methods) perform the same amount of computation each time.

If a complete optimal value function is known, a globally optimal solution follows simply by greedily choosing actions to optimize it. However, value-function based approaches have thus far been difficult to translate into high dimensional robotics as they require function approximation for the value function. Most theoretical guarantees no longer hold for this approximation and even finding the optimal action can be a hard problem due to the brittleness of the approximation and the cost of optimization. For high dimensional actions, it can be as hard computing an improved policy for all states in policy search as finding a single optimal action on-policy for one state by searching the state-action value function.

In principle, a value function requires total coverage of the state space and the *largest local error* determines the quality of the resulting policy. A particularly significant problem is the error propagation in value functions.

A small change in the policy may cause a large change in the value function, which again causes a large change in the policy. While this may lead more quickly to good, possibly globally optimal solutions, such learning processes often prove unstable under function approximation (Boyan and Moore, 1994; Kakade and Langford, 2002; Bagnell et al, 2003) and are considerably more dangerous when applied to real systems where overly large policy deviations may lead to dangerous decisions.

In contrast, policy search methods usually only consider the current policy and its neighborhood in order to gradually improve performance. The result is that usually only local optima, and not the global one, can be found. However, these methods work well in conjunction with continuous features. Local coverage and local errors results into improved scaleability in robotics.

Policy search methods are sometimes called *actor*-only methods; value function methods are sometimes called *critic*-only methods. The idea of a critic is to first observe and estimate the performance of choosing controls on the system (i.e., the value function), then derive a policy based on the gained knowledge. In contrast, the actor directly tries to deduce the optimal policy. A set of algorithms called *actor-critic* methods attempt to incorporate the advantages of each: a policy is explicitly maintained, as is a value-function for the current policy. The value function (i.e., the critic) is not employed for action selection. Instead, it observes the performance of the actor and decides when the policy needs to be updated and which action should be preferred. The resulting update step features the local convergence properties of policy gradient algorithms while reducing update variance (Greensmith et al, 2004). There is a trade-off between the benefit of reducing the variance of the updates and having to learn a value function as the samples required to estimate the value function could also be employed to obtain better gradient estimates for the update step. Rosenstein and Barto (2004) propose an actor-critic method that additionally features a supervisor in the form of a stable policy.

2.2.4   Function Approximation

Function approximation (Rivlin, 1969) is a family of mathematical and statistical techniques used to represent a function of interest when it is computationally or information-theoretically intractable to represent the function exactly or explicitly (e.g. in tabular form). Typically, in reinforcement learning te function approximation is based on sample data collected during interaction with the environment. Function approximation is critical in nearly every RL problem, and becomes inevitable in continuous state ones. In large discrete spaces it is also often impractical to visit or even represent all states and actions, and function approximation in this setting can be used as a means to generalize to neighboring states and actions.

Function approximation can be employed to represent policies, value functions, and forward models. Broadly speaking, there are two kinds of function approximation methods: *parametric* and *non-parametric*. A parametric function approximator uses a finite set of parameters or arguments with the goal is to find parameters that make this approximation fit the observed data as closely as possible. Examples include linear basis functions and neural networks. In contrast, non-parametric methods expand representational power in relation to collected data and hence are not limited by the representation power of a chosen parametrization (Bishop, 2006). A prominent example that has found much use within reinforcement learning is Gaussian process regression (Rasmussen and Williams, 2006). A fundamental problem with using supervised learning methods developed in the literature for function approximation is that most such methods are designed for independently and identically distributed sample data. However, the data generated by the reinforcement learning process is usually neither independent nor identically distributed. Usually, the function approximator itself plays some role in the data collection process (for instance, by serving to define a policy that we execute on a robot.)

Linear basis function approximators form one of the most widely used approximate value function techniques in continuous (and discrete) state spaces. This is largely due to the simplicity of their representation as well as a convergence theory, albeit limited, for the approximation of value functions based on samples (Tsitsiklis and Van Roy, 1997). Let us briefly take a closer look at a radial basis function network to illustrate this approach. The value function maps states to a scalar value. The state space can be covered by a grid of points, each of which correspond to the center of a Gaussian-shaped basis function. The value of the approximated function is the weighted sum of the values of all basis functions at the query point. As the influence of the Gaussian basis functions drops rapidly, the value of the query points will be predominantly influenced by the neighboring basis functions. The weights are set in a way to minimize the error between the observed samples and the reconstruction. For the mean squared error, these weights can be determined by linear regression. Kolter and Ng (2009b) discuss the benefits of regularization of such linear function approximators to avoid over-fitting.

Other possible function approximators for value functions include wire fitting, whichBaird and Klopf (1993) suggested as an approach that makes continuous action selection feasible. The Fourier basis had been suggested by Konidaris et al (2011b). Even discretizing the state-space can be seen as a form of function approximation where coarse values serve as estimates for a smooth continuous function. One example is tile coding (Sutton and Barto, 1998), where the space is subdivided into (potentially irregularly shaped) regions, called tiling. The number of different tilings determines the resolution of the final approximation. For more examples, please refer to Sections 2.4.1 and 2.4.2.

Policy search also benefits from a compact representation of the policy as discussed in Section 2.4.3.

Models of the system dynamics can be represented using a wide variety of techniques. In this case, it is often important to model the uncertainty in the model (e.g., by a stochastic model or Bayesian estimates of model parameters) to ensure that the learning algorithm does not exploit model inaccuracies. See Section 2.6 for a more detailed discussion.

2.3   Challenges in Robot Reinforcement Learning

Reinforcement learning is generally a hard problem and many of its challenges are particularly apparent in the robotics setting. As the states and actions of most robots are inherently continuous, we are forced to consider the resolution at which they are represented. We must decide how fine grained the control is that we require over the robot, whether we employ discretization or function approximation, and what time step we establish. Additionally, as the dimensionality of both states and actions can be high, we face the "Curse of Dimensionality" (Bellman, 1957) as discussed in Section 2.3.1. As robotics deals with complex physical systems, samples can be expensive due to the long execution time of complete tasks, required manual interventions, and the need maintenance and repair. In these real-world measurements, we must cope with the uncertainty inherent in complex physical systems. A robot requires that the algorithm runs in real-time. The algorithm must be capable of dealing with delays in sensing and execution that are inherent in physical systems (see Section 2.3.2). A simulation might alleviate many problems but these approaches need to be robust with respect to model errors as discussed in Section 2.3.3. An often underestimated problem is the goal specification, which is achieved by designing a good reward function. As noted in Section 2.3.4, this choice can make the difference between feasibility and an unreasonable amount of exploration.

2.3.1   Curse of Dimensionality

When Bellman (1957) explored optimal control in discrete high-dimensional spaces, he faced an exponential explosion of states and actions for which he coined the term "Curse of Dimensionality". As the number of dimensions grows, exponentially more data and computation are needed to cover the complete state-action space. For example, if we assume that each dimension of a state-space is discretized into ten levels, we have 10 states for a one-dimensional state-space, $10^3 = 1000$ unique states for a three-dimensional state-space, and 10^n possible states for a n-dimensional state space. Evaluating every state quickly becomes infeasible with growing dimensionality, even for discrete states. Bellman originally coined the term in the context of optimization, but it also applies to function approximation and numerical integration (Donoho, 2000). While supervised learning methods have tamed

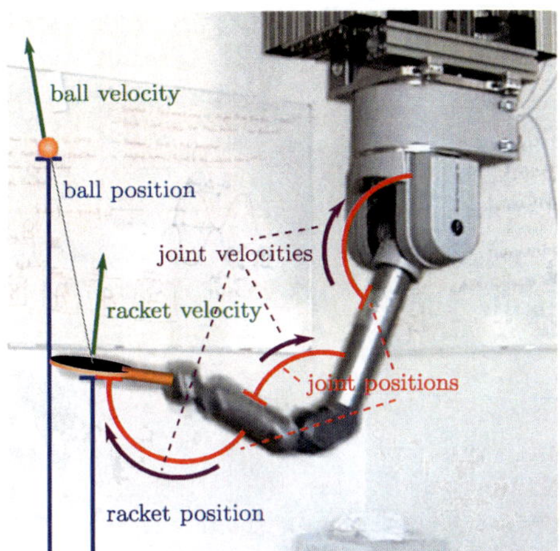

Fig. 2.3 This Figure illustrates the state space used in the modeling of a robot reinforcement learning task of paddling a ball

this exponential growth by considering only competitive optimality with respect to a limited class of function approximators, such results are much more difficult in reinforcement learning where data must collected throughout state-space to ensure global optimality.

Robotic systems often have to deal with these high dimensional states and actions due to the many degrees of freedom of modern anthropomorphic robots. For example, in the ball-paddling task shown in Figure 2.3, a proper representation of a robot's state would consist of its joint angles and velocities for each of its seven degrees of freedom as well as the Cartesian position and velocity of the ball. The robot's actions would be the generated motor commands, which often are torques or accelerations. In this example, we have $2 \times (7 + 3) = 20$ state dimensions and 7-dimensional continuous actions. Obviously, other tasks may require even more dimensions. For example, human-like actuation often follows the antagonistic principle (Yamaguchi and Takanishi, 1997) which additionally enables control of stiffness. Such dimensionality is a major challenge for both the robotics and the reinforcement learning communities.

In robotics, such tasks are often rendered tractable to the robot engineer by a hierarchical task decomposition that shifts some complexity to a lower layer of functionality. Classical reinforcement learning approaches often consider a grid-based representation with discrete states and actions, often referred to as a *grid-world*. A navigational task for mobile robots could be projected into this representation by employing a number of actions like "move to the cell to the left" that use a lower level controller that takes care of accelerating, moving, and stopping while ensuring precision. In the ball-paddling example, we may simplify by controlling the robot in racket space (which is lower-dimensional as the racket is orientation-invariant around the string's mounting point) with an operational space control law (Nakanishi et al, 2008). Many commercial robot systems also encapsulate some of the state and action components in an embedded control system (e.g., trajectory fragments are frequently used as actions for industrial robots). However, this form of a state dimensionality reduction severely limits the dynamic capabilities of the robot according to our experience (Schaal et al, 2002; Peters et al, 2010b).

The reinforcement learning community has a long history of dealing with dimensionality using computational abstractions. It offers a larger set of applicable tools ranging from adaptive discretizations (Buşoniu et al, 2010) and function approximation approaches (Sutton and Barto, 1998) to macro-actions or options (Barto and Mahadevan, 2003; Hart and Grupen, 2011). Options allow a task to be decomposed into elementary components and quite naturally translate to robotics. Such options can autonomously achieve a sub-task, such as opening a door, which reduces the planning horizon (Barto and Mahadevan, 2003). The automatic generation of such sets of options is a key issue in order to enable such approaches. We will discuss approaches that have been successful in robot reinforcement learning in Section 2.4.

2.3.2 Curse of Real-World Samples

Robots inherently interact with the physical world. Hence, robot reinforcement learning suffers from most of the resulting real-world problems. For example, robot hardware is usually expensive, suffers from wear and tear, and requires careful maintenance. Repairing a robot system is a non-negligible effort associated with cost, physical labor and long waiting periods. To apply reinforcement learning in robotics, safe exploration becomes a key issue of the learning process (Schneider, 1996; Bagnell, 2004; Deisenroth and Rasmussen, 2011; Moldovan and Abbeel, 2012), a problem often neglected in the general reinforcement learning community. Perkins and Barto (2002) have come up with a method for constructing reinforcement learning agents based on Lyapunov functions. Switching between the underlying controllers is always safe and offers basic performance guarantees.

However, several more aspects of the real-world make robotics a challenging domain. As the dynamics of a robot can change due to many external factors ranging from temperature to wear, the learning process may never fully converge, i.e., it needs a "tracking solution" (Sutton et al, 2007). Frequently, the environment settings during an earlier learning period cannot be reproduced. External factors are not always clear – for example, how light conditions affect the performance of the vision system and, as a result, the task's performance. This problem makes comparing algorithms particularly hard. Furthermore, the approaches often have to deal with uncertainty due to inherent measurement noise and the inability to observe all states directly with sensors.

Most real robot learning tasks require some form of human supervision, e.g., putting the pole back on the robot's end-effector during pole balancing (see Figure 2.1d) after a failure. Even when an automatic reset exists (e.g., by having a smart mechanism that resets the pole), learning speed becomes essential as a task on a real robot cannot be sped up. In some tasks like a slowly rolling robot, the dynamics can be ignored; in others like a flying robot, they cannot. Especially in the latter case, often the whole episode needs to be completed as it is not possible to start from arbitrary states.

For such reasons, real-world samples are expensive in terms of time, labor and, potentially, finances. In robotic reinforcement learning, it is often considered to be more important to limit the real-world interaction time instead of limiting memory consumption or computational complexity. Thus, sample efficient algorithms that are able to learn from a small number of trials are essential. In Section 2.6 we will point out several approaches that allow the amount of required real-world interactions to be reduced.

Since the robot is a physical system, there are strict constraints on the interaction between the learning algorithm and the robot setup. For dynamic tasks, the movement cannot be paused and actions must be selected within a time-budget without the opportunity to pause to think, learn or plan between actions. These constraints are less severe in an episodic setting where the time intensive part of the learning can be postponed to the period between episodes. Hester et al (2012) has proposed a real-time architecture for model-based value function reinforcement learning methods taking into account these challenges.

As reinforcement learning algorithms are inherently implemented on a digital computer, the discretization of time is unavoidable despite that physical systems are inherently continuous time systems. Time-discretization of the actuation can generate undesirable artifacts (e.g., the distortion of distance between states) even for idealized physical systems, which cannot be avoided. As most robots are controlled at fixed sampling frequencies (in the range between 500Hz and 3kHz) determined by the manufacturer of the robot, the upper bound on the rate of temporal discretization is usually pre-determined. The lower bound depends on the horizon of the problem, the achievable speed of changes in the state, as well as delays in sensing and actuation.

All physical systems exhibit such delays in sensing and actuation. The state of the setup (represented by the filtered sensor signals) may frequently lag behind the real state due to processing and communication delays. More critically, there are also communication delays in actuation as well as delays due to the fact that neither motors, gear boxes nor the body's movement can change instantly. Due to these delays, actions may not have instantaneous effects but are observable only several time steps later. In contrast, in most general reinforcement learning algorithms, the actions are assumed to take effect instantaneously as such delays would violate the usual Markov assumption. This effect can be addressed by putting some number of recent actions into the state. However, this significantly increases the dimensionality of the problem.

The problems related to time-budgets and delays can also be avoided by increasing the duration of the time steps. One downside of this approach is that the robot cannot be controlled as precisely; another is that it may complicate a description of system dynamics.

2.3.3 Curse of Under-Modeling and Model Uncertainty

One way to offset the cost of real-world interaction is to use accurate models as simulators. In an ideal setting, this approach would render it possible to learn the behavior in simulation and subsequently transfer it to the real robot. Unfortunately, creating a sufficiently accurate model of the robot and its environment is challenging and often requires very many data samples. As small model errors due to this under-modeling accumulate, the simulated robot can quickly diverge from the real-world system. When a policy is trained using an imprecise forward model as simulator, the behavior will not transfer without significant modifications as experienced by Atkeson (1994) when learning the underactuated pendulum swing-up. The authors have achieved a direct transfer in only a limited number of experiments; see Section 2.6.1 for examples.

For tasks where the system is self-stabilizing (that is, where the robot does not require active control to remain in a safe state or return to it), transferring policies often works well. Such tasks often feature some type of dampening that absorbs the energy introduced by perturbations or control inaccuracies. If the task is inherently stable, it is safer to assume that approaches that were applied in simulation work similarly in the real world (Kober and Peters, 2011b). Nevertheless, tasks can often be learned better in the real world than in simulation due to complex mechanical interactions (including contacts and friction) that have proven difficult to model accurately. For example, in the ball-paddling task (Figure 2.3) the elastic string that attaches the ball to the racket always pulls back the ball towards the racket even when hit very hard. Initial simulations (including friction models, restitution models, dampening models, models for the elastic string, and air drag) of the ball-racket contacts indicated that these factors would be very hard to control. In a real experiment, however, the reflections of the ball on the racket proved to be less critical than in simulation and the stabilizing forces due to the elastic string were sufficient to render the whole system self-stabilizing.

In contrast, in unstable tasks small variations have drastic consequences. For example, in a pole balancing task, the equilibrium of the upright pole is very brittle and constant control is required to stabilize the system. Transferred policies often perform poorly in this setting. Nevertheless, approximate models serve a number of key roles which we discuss in Section 2.6, including verifying and testing the algorithms in simulation, establishing proximity to theoretically optimal solutions, calculating approximate gradients for local policy improvement, identifing strategies for collecting more data, and performing "mental rehearsal".

2.3.4   Curse of Goal Specification

In reinforcement learning, the desired behavior is implicitly specified by the reward function. The goal of reinforcement learning algorithms then is to maximize the accumulated long-term reward. While often dramatically simpler than specifying the behavior itself, in practice, it can be surprisingly difficult to define a good reward function in robot reinforcement learning. The learner must observe variance in the reward signal in order to be able to improve a policy: if the same return is always received, there is no way to determine which policy is better or closer to the optimum.

In many domains, it seems natural to provide rewards only upon task achievement – for example, when a table tennis robot wins a match. This view results in an apparently simple, binary reward specification. However, a robot may receive such a reward so rarely that it is unlikely to ever succeed in the lifetime of a real-world system. Instead of relying on simpler binary rewards, we frequently need to include intermediate rewards in the scalar reward function to guide the learning process to a reasonable solution, a process known as *reward shaping* (Laud, 2004).

Beyond the need to shorten the effective problem horizon by providing intermediate rewards, the trade-off between different factors may be essential. For instance, hitting a table tennis ball very hard may result in a high score but is likely to damage a robot or shorten its life span. Similarly, changes in actions may be penalized to avoid high frequency controls that are likely to be very poorly captured with tractable low dimensional state-space or rigid-body models. Reinforcement learning algorithms are also notorious for exploiting the reward function in ways that are not anticipated by the designer. For example, if the distance between the ball and the desired highest point is part of the reward in ball paddling (see Figure 2.3), many locally optimal solutions would attempt to simply move the racket upwards and keep the ball on it. Reward shaping gives the system a notion of closeness to the desired behavior instead of relying on a reward that only encodes success or failure (Ng et al, 1999).

Often the desired behavior can be most naturally represented with a reward function in a particular state and action space. However, this representation does not necessarily correspond to the space where the actual learning needs to be performed due to both computational and statistical limitations. Employing methods to render the learning problem tractable often result in different, more abstract state and action spaces which might not allow accurate representation of the original reward function. In such cases, a reward artfully specified in terms of the features of the space in which the learning algorithm operates can prove remarkably effective. There is also a trade-off between the complexity of the reward function and the complexity of the learning problem. For example, in the ball-in-a-cup task (Section 2.7) the most natural reward would be a binary value depending on whether the ball is in the cup or not. To render the learning problem tractable, a less intuitive reward needed to be devised in terms of a Cartesian distance with additional directional information (see Section 2.7.1 for details). Another example is Crusher (Ratliff et al, 2006a), an outdoor robot, where the human designer was interested in a combination of minimizing time and risk to the robot. However, the robot reasons about the world on the long time horizon scale as if it was a very simple, deterministic, holonomic robot operating on a fine grid of continuous costs. Hence, the desired behavior cannot be represented straightforwardly in this state-space. Nevertheless, a remarkably human-like behavior that seems to respect time and risk priorities can be achieved by carefully mapping features describing each state (discrete grid location with features computed by an on-board perception system) to cost.

Inverse optimal control, also known as inverse reinforcement learning (Russell, 1998), is a promising alternative to specifying the reward function manually. It assumes that a reward function can be reconstructed from a set of expert demonstrations. This reward function does not necessarily correspond to the true reward function, but provides guarantees on the resulting performance of learned behaviors (Abbeel and Ng, 2004; Ratliff et al, 2006b). Inverse optimal control was initially studied in the control community (Kalman, 1964) and in the field of economics (Keeney and Raiffa, 1976). The initial results were only applicable to limited domains (linear quadratic regulator problems) and required closed form access to plant and controller, hence samples from human demonstrations could not be used. Russell (1998) brought the field to the attention of the machine learning community. Abbeel and Ng (2004) defined an important constraint on the solution to the inverse RL problem when reward functions are linear in a set of features: a policy that is extracted by observing demonstrations has to earn the same reward as the policy that is being demonstrated. Ratliff et al (2006b) demonstrated that inverse optimal control can be understood as a generalization of ideas in machine learning of *structured prediction* and introduced efficient sub-gradient based algorithms with regret bounds that enabled large scale application of the technique within robotics. Ziebart et al (2008) extended the technique developed by Abbeel and Ng (2004) by rendering the idea robust and probabilistic, enabling its effective use for both learning policies and predicting the behavior of sub-optimal agents. These techniques, and many variants, have been recently successfully applied to outdoor robot navigation (Ratliff et al, 2006a; Silver et al, 2008, 2010), manipulation (Ratliff et al, 2007), and quadruped locomotion (Ratliff et al, 2006a, 2007; Kolter et al, 2007).

More recently, the notion that complex policies can be built on top of simple, easily solved optimal control problems by exploiting rich, parametrized reward functions has been exploited within reinforcement learning more directly. In (Sorg et al, 2010; Zucker and Bagnell, 2012), complex policies are derived by adapting a reward function for simple optimal control problems using policy search techniques. Zucker and Bagnell (2012) demonstrate that this technique can enable efficient solutions to robotic marble-maze problems that effectively transfer between mazes of varying design and complexity.

These works highlight the natural trade-off between the complexity of the reward function and the complexity of the underlying reinforcement learning problem for achieving a desired behavior.

2.4   Tractability through Representation

As discussed above, reinforcement learning provides a framework for a remarkable variety of problems of significance to both robotics and machine learning. However, the computational and information-theoretic consequences that we outlined above accompany this power and generality. As a result, naive application of reinforcement learning techniques in robotics is likely to be doomed to failure. The remarkable successes that we reference in this article have been achieved by leveraging a few key principles – effective representations, approximate models, and prior knowledge or information. In the following three sections, we review these principles and summarize how each has been made effective in practice. We hope that understanding these broad approaches will lead to new successes in robotic reinforcement learning by combining successful methods and encourage research on novel techniques that embody each of these principles.

Much of the success of reinforcement learning methods has been due to the clever use of approximate representations. The need of such approximations is particularly pronounced in robotics, where table-based representations (as discussed in Section 2.2.2) are rarely scalable. The different ways of making reinforcement learning methods tractable in robotics are tightly coupled to the underlying optimization framework. Reducing the dimensionality of states or actions by smart state-action discretization is a representational simplification that may enhance both policy search and value function-based methods (see Section 2.4.1). A value function-based approach requires an accurate and robust but general function approximator that can capture the value function with sufficient precision (see Section 2.4.2) while maintaining stability during learning. Policy search methods require a choice of policy representation that controls the complexity of representable policies to enhance learning speed (see Section 2.4.3). An overview of publications that make particular use of efficient representations to render the learning problem tractable is presented in Tables 2.3.

2.4.1   Smart State-Action Discretization

Decreasing the dimensionality of state or action spaces eases most reinforcement learning problems significantly, particularly in the context of robotics. Here, we give a short overview of different attempts to achieve this goal with smart discretization.

Hand Crafted Discretization. A variety of authors have manually developed discretizations so that basic tasks can be learned on real robots. For

Table 2.3 This table illustrates different methods of making robot reinforcement learning tractable by employing a suitable representation

SMART STATE-ACTION DISCRETIZATION

Approach	Employed by...
Hand crafted	Benbrahim et al (1992); Kimura et al (2001); Kwok and Fox (2004); Nemec et al (2010); Paletta et al (2007); Tokic et al (2009); Willgoss and Iqbal (1999)
Learned	Piater et al (2011); Yasuda and Ohkura (2008)
Meta-actions	Asada et al (1996); Dorigo and Colombetti (1993); Fidelman and Stone (2004); Huber and Grupen (1997); Kalmár et al (1998); Konidaris et al (2011a, 2012); Matarić (1994, 1997); Platt et al (2006); Soni and Singh (2006); Nemec et al (2009)
Relational Representation	Cocora et al (2006); Katz et al (2008)

VALUE FUNCTION APPROXIMATION

Approach	Employed by...
Physics-inspired Features	An et al (1988); Schaal (1996)
Neural Networks	Benbrahim and Franklin (1997); Duan et al (2008); Gaskett et al (2000); Hafner and Riedmiller (2003); Riedmiller et al (2009); Thrun (1995)
Neighbors	Hester et al (2010); Mahadevan and Connell (1992); Touzet (1997)
Local Models	Bentivegna (2004); Schaal (1996); Smart and Kaelbling (1998)
GPR	Gräve et al (2010); Kroemer et al (2009, 2010); Rottmann et al (2007)

PRE-STRUCTURED POLICIES

Approach	Employed by...
Via Points & Splines	Kuindersma et al (2011); Miyamoto et al (1996); Roberts et al (2010)
Linear Models	Tamei and Shibata (2009)
Motor Primitives	Kohl and Stone (2004); Kober and Peters (2008); Peters and Schaal (2008c,b); Stulp et al (2011); Tamošiūnaitė et al (2011); Theodorou et al (2010)
GMM & RBF	Deisenroth and Rasmussen (2011); Deisenroth et al (2011); Guenter et al (2007); Lin and Lai (2012); Peters and Schaal (2008a)
Neural Networks	Endo et al (2008); Geng et al (2006); Gullapalli et al (1994); Hailu and Sommer (1998); Bagnell and Schneider (2001)
Controllers	Bagnell and Schneider (2001); Kolter and Ng (2009a); Tedrake (2004); Tedrake et al (2005); Vlassis et al (2009); Zucker and Bagnell (2012)
Non-parametric	Kober et al (2010b); Mitsunaga et al (2005); Peters et al (2010a)

low-dimensional tasks, we can generate discretizations straightforwardly by splitting each dimension into a number of regions. The main challenge is to find the right number of regions for each dimension that allows the system to achieve a good final performance while still learning quickly. Example applications include balancing a ball on a beam (Benbrahim et al, 1992), one degree of freedom ball-in-a-cup (Nemec et al, 2010), two degree of freedom crawling motions (Tokic et al, 2009), and gait patterns for four legged walking (Kimura et al, 2001). Much more human experience is needed for more complex tasks. For example, in a basic navigation task with noisy sensors (Willgoss and Iqbal, 1999), only some combinations of binary state or action indicators are useful (e.g., you can drive left and forward at the same time, but not backward and forward). The state space can also be based on vastly different features, such as positions, shapes, and colors, when learning object affordances (Paletta et al, 2007) where both the discrete sets and the mapping from sensor values to the discrete values need to be crafted. Kwok and Fox (2004) use a mixed discrete and continuous representation of the state space to learn active sensing strategies in a RoboCup scenario. They first discretize the state space along the dimension with the strongest non-linear influence on the value function and subsequently employ a linear value function approximation (Section 2.4.2) for each of the regions.

Learned from Data. Instead of specifying the discretizations by hand, they can also be built adaptively during the learning process. For example, a rule based reinforcement learning approach automatically segmented the state space to learn a cooperative task with mobile robots (Yasuda and Ohkura, 2008). Each rule is responsible for a local region of the state-space. The importance of the rules are updated based on the rewards and irrelevant rules are discarded. If the state is not covered by a rule yet, a new one is added. In the related field of computer vision, Piater et al (2011) propose an approach that adaptively and incrementally discretizes a perceptual space into discrete states, training an image classifier based on the experience of the RL agent to distinguish visual classes, which correspond to the states.

Meta-Actions. Automatic construction of meta-actions (and the closely related concept of options) has fascinated reinforcement learning researchers and there are various examples in the literature. The idea is to have more intelligent actions that are composed of a sequence of movements and that in themselves achieve a simple task. A simple example would be to have a meta-action "move forward 5m." A lower level system takes care of accelerating, stopping, and correcting errors. For example, in (Asada et al, 1996), the state and action sets are constructed in a way that repeated action primitives lead to a change in the state to overcome problems associated with the discretization. Q-learning and dynamic programming based approaches have been compared in a pick-n-place task (Kalmár et al, 1998) using modules. Huber and Grupen (1997) use a set of controllers with associated predicate states as a basis for learning turning gates with a quadruped. Fidelman and Stone (2004) use a policy search approach to learn a small set of parameters that controls the transition between a walking and a capturing meta-action in a RoboCup scenario. A task of transporting a ball with a dog robot (Soni and Singh, 2006) can be learned with semi-automatically discovered options. Using only the sub-goals of primitive motions, a humanoid robot can learn a pouring task (Nemec et al, 2009). Other examples include foraging (Matarić, 1997) and cooperative tasks (Matarić, 1994) with multiple robots, grasping with restricted search spaces (Platt et al, 2006), and mobile robot navigation (Dorigo and Colombetti, 1993). If the meta-actions are not fixed in advance, but rather learned at the same time, these approaches are hierarchical reinforcement learning approaches as discussed in Section 2.5.2. Konidaris et al (2011a, 2012) propose an approach that constructs a skill tree from human demonstrations. Here, the skills correspond to options and are chained to learn a mobile manipulation skill.

Relational Representations. In a relational representation, the states, actions, and transitions are not represented individually. Entities of the same predefined type are grouped and their relationships are considered. This representation may be preferable for highly geometric tasks (which frequently appear in robotics) and has been employed to learn to navigate buildings with a real robot in a supervised setting (Cocora et al, 2006) and to manipulate articulated objects in simulation (Katz et al, 2008).

2.4.2   Value Function Approximation

Function approximation has always been the key component that allowed value function methods to scale into interesting domains. In robot reinforcement learning, the following function approximation schemes have been popular and successful. Using function approximation for the value function can be combined with using function approximation for learning a model of the system (as discussed in Section 2.6) in the case of model-based reinforcement learning approaches.

Unfortunately the max-operator used within the Bellman equation and temporal-difference updates can theoretically make most linear or non-linear approximation schemes unstable for either value iteration or policy iteration. Quite frequently such an unstable behavior is also exhibited in practice. Linear function approximators are stable for policy evaluation, while non-linear function approximation (e.g., neural networks) can even diverge if just used for policy evaluation (Tsitsiklis and Van Roy, 1997).

Physics-inspired Features. If good hand-crafted features are known, value function approximation can be accomplished using a linear combination of features. However, good features are well known in robotics only for a few problems, such as features for local stabilization (Schaal, 1996) and features describing rigid body dynamics (An et al, 1988). Stabilizing a system at an unstable equilibrium point is the most well-known example, where a second order Taylor expansion of the state together with a linear value function approximator often suffice as features in the proximity of the equilibrium point. For example, Schaal (1996) showed that such features suffice for learning how to stabilize a pole on the end-effector of a robot when within $\pm 15 - 30$ degrees of the equilibrium angle. For sufficient features, linear function approximation is likely to yield good results in an on-policy setting. Nevertheless, it is straightforward to show that impoverished value function representations

Fig. 2.4 The Brainstormer Tribots won the RoboCup 2006 MidSize League (Riedmiller et al, 2009)(Picture reprint with permission of Martin Riedmiller)

(e.g., omitting the cross-terms in quadratic expansion in Schaal's set-up) will make it impossible for the robot to learn this behavior. Similarly, it is well known that linear value function approximation is unstable in the off-policy case (Tsitsiklis and Van Roy, 1997; Gordon, 1999; Sutton and Barto, 1998).

Neural Networks. As good hand-crafted features are rarely available, various groups have employed neural networks as global, non-linear value function approximation. Many different flavors of neural networks have been applied in robotic reinforcement learning. For example, multi-layer perceptrons were used to learn a wandering behavior and visual servoing (Gaskett et al, 2000). Fuzzy neural networks (Duan et al, 2008) and explanation-based neural networks (Thrun, 1995) have allowed robots to learn basic navigation. CMAC neural networks have been used for biped locomotion (Benbrahim and Franklin, 1997).

The Brainstormers RoboCup soccer team is a particularly impressive application of value function approximation.(see Figure 2.4). It used multi-layer perceptrons to learn various sub-tasks such as learning defenses, interception, position control, kicking, motor speed control, dribbling and penalty shots (Hafner and Riedmiller, 2003; Riedmiller et al, 2009). The resulting components contributed substantially to winning the world cup several times in the simulation and the mid-size real robot leagues. As neural networks are global function approximators, overestimating the value function at a frequently occurring state will increase the values predicted by the neural network for all other states, causing fast divergence (Boyan and Moore, 1994; Gordon, 1999). Riedmiller et al (2009) solved this problem by always defining an absorbing state where they set the value predicted by their neural network to zero, which "clamps the neural network down" and thereby prevents divergence. It also allows re-iterating on the data, which results in an improved value function quality. The combination of iteration on data with the clamping technique appears to be the key to achieving good performance with value function approximation in practice.

Generalize to Neighboring Cells. As neural networks are globally affected from local errors, much work has focused on simply generalizing from neighboring cells. One of the earliest papers in robot reinforcement learning (Mahadevan and Connell, 1992) introduced this idea by statistically clustering states to speed up a box-pushing task with a mobile robot, see Figure 2.1a. This approach was also used for a navigation and obstacle avoidance task with a mobile robot (Touzet, 1997). Similarly, decision trees have been used to generalize states and actions to unseen ones, e.g., to learn a penalty kick on a humanoid robot (Hester et al, 2010). The core problem of these methods is the lack of scalability to high-dimensional state and action spaces.

Local Models. Local models can be seen as an extension of generalization among neighboring cells to generalizing among neighboring data points. Locally weighted regression creates particularly efficient function approximation in the context of robotics both in supervised and reinforcement learning. Here, regression errors are weighted down by proximity to query point to train local modelsThe predictions of these local models are combined using the same weighting functions. Using local models for value function approximation has allowed learning a navigation task with obstacle avoidance (Smart and Kaelbling, 1998), a pole swing-up task (Schaal, 1996), and an air hockey task (Bentivegna, 2004).

Gaussian Process Regression. Parametrized global or local models need to pre-specify, which requires a trade-off between representational accuracy and the number of parameters. A non-parametric function approximator like Gaussian Process Regression (GPR) could be employed instead, but potentially at the cost of a higher computational complexity. GPR has the added advantage of providing a notion of uncertainty about the approximation quality for a query point. Hovering with an autonomous blimp (Rottmann et al, 2007) has been achieved by approximation the state-action value function with a GPR. Similarly, another paper shows that grasping can be learned using Gaussian process regression (Gräve et al, 2010) by additionally taking into account the uncertainty to guide the exploration. Grasping locations can be learned by approximating the rewards with a GPR, and trying candidates with predicted high rewards (Kroemer et al, 2009), resulting in an active learning approach. High reward uncertainty allows intelligent exploration in reward-based grasping (Kroemer et al, 2010) in a bandit setting.

2.4.3   Pre-structured Policies

Policy search methods greatly benefit from employing an appropriate function approximation of the policy. For example, when employing gradient-based approaches, the trade-off between the representational power of the policy (in the form of many policy parameters) and the learning speed (related to the number of samples required to estimate the gradient) needs to be considered. To make policy search approaches tractable, the policy needs to be represented

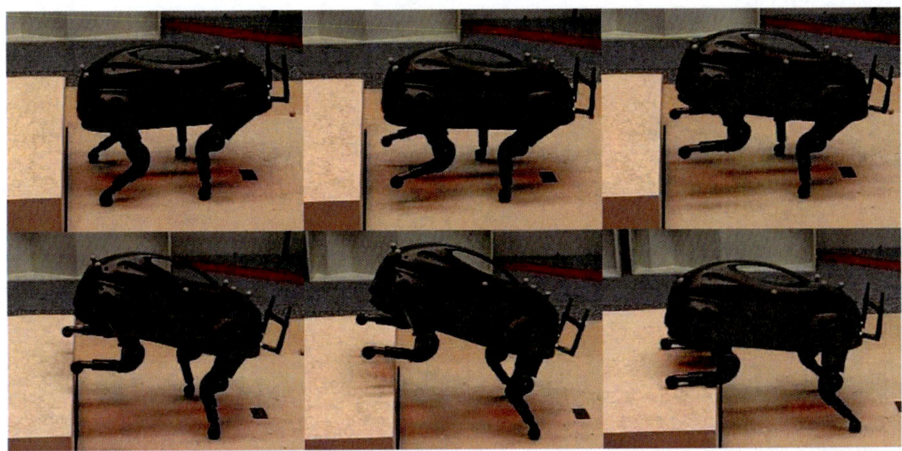

Fig. 2.5 Boston Dynamics LittleDog jumping (Kolter and Ng, 2009a) (Picture reprint with permission of Zico Kolter)

with a function approximation that takes into account domain knowledge, such as task-relevant parameters or generalization properties. As the next action picked by a policy depends on the current state and action, a policy can be seen as a closed-loop controller. Roberts et al (2011) demonstrate that care needs to be taken when selecting closed-loop parameterizations for weakly-stable systems, and suggest forms that are particularly robust during learning. However, especially for episodic RL tasks, sometimes open-loop policies (i.e., policies where the actions depend only on the time) can also be employed.

Via Points and Splines. An open-loop policy may often be naturally represented as a trajectory, either in the space of states or targets or directly as a set of controls. Here, the actions are only a function of time, which can be considered as a component of the state. Such spline-based policies are very suitable for compressing complex trajectories into few parameters. Typically the desired joint or Cartesian position, velocities, and/or accelerations are used as actions. To minimize the required number of parameters, not every point is stored. Instead, only important via-points are considered and other points are interpolated. Miyamoto et al (1996) optimized the position and timing of such via-points in order to learn a kendama task (a traditional Japanese toy similar to ball-in-a-cup). A well known type of a via point representations are splines, which rely on piecewise-defined smooth polynomial functions for interpolation. For example, Roberts et al (2010) used a periodic cubic spline as a policy parametrization for a flapping system and Kuindersma et al (2011) used a cubic spline to represent arm movements in an impact recovery task.

Linear Models. If model knowledge of the system is available, it can be used to create features for linear closed-loop policy representations. For example,

Tamei and Shibata (2009) used policy-gradient reinforcement learning to adjust a model that maps from human EMG signals to forces that in turn is used in a cooperative holding task.

Motor Primitives. Motor primitives combine linear models describing dynamics with parsimonious movement parametrizations. While originally biologically-inspired, they have a lot of success for representing basic movements in robotics such as a reaching movement or basic locomotion. These basic movements can subsequently be sequenced and/or combined to achieve more complex movements. For both goal oriented and rhythmic movement, different technical representations have been proposed in the robotics community. Dynamical systems motor primitives (Ijspeert et al, 2002a; Schaal et al, 2007) have become a popular representation for reinforcement learning of discrete movements. The dynamical systems motor primitives always have a strong dependence on the phase of the movement, which corresponds to time. They can be employed as an open-loop trajectory representation. Nevertheless, they can also be employed as a closed-loop policy to a limited extent. In our experience, they offer a number of advantages over via-point or spline based policy representation (see Section 2.7.2). The dynamical systems motor primitives have been trained with reinforcement learning for a T-ball batting task (Peters and Schaal, 2008c,b), an underactuated pendulum swing-up and a ball-in-a-cup task (Kober and Peters, 2008), flipping a light switch (Buchli et al, 2011), pouring water (Tamošiūnaitė et al, 2011), and playing pool and manipulating a box (Pastor et al, 2011). For rhythmic behaviors, a representation based on the same biological motivation but with a fairly different technical implementation (based on half-elliptical locuses) have been used to acquire the gait patterns for an Aibo robot dog locomotion (Kohl and Stone, 2004).

Gaussian Mixture Models and Radial Basis Function Models. When more general policies with a strong state-dependence are needed, general function approximators based on radial basis functions, also called Gaussian kernels, become reasonable choices. While learning with fixed basis function centers and widths often works well in practice, estimating them is challenging. These centers and widths can also be estimated from data prior to the reinforcement learning process. This approach has been used to generalize a open-loop reaching movement (Guenter et al, 2007; Lin and Lai, 2012) and to learn the closed-loop cart-pole swingup task (Deisenroth and Rasmussen, 2011). Globally linear models were employed in a closed-loop block stacking task (Deisenroth et al, 2011).

Neural Networks are another general function approximation used to represent policies. Neural oscillators with sensor feedback have been used to learn rhythmic movements where open and closed-loop information were combined, such as gaits for a two legged robot (Geng et al, 2006; Endo et al, 2008). Similarly, a peg-in-hole (see Figure 2.1b), a ball-balancing task (Gullapalli et al, 1994), and a navigation task (Hailu and Sommer, 1998) have been learned with closed-loop neural networks as policy function approximators.

Locally Linear Controllers. As local linearity is highly desirable in robot movement generation to avoid actuation difficulties, learning the parameters of a locally linear controller can be a better choice than using a neural network or radial basis function representation. Several of these controllers can be combined to form a global, inherently closed-loop policy. This type of policy has allowed for many applications, including learning helicopter flight (Bagnell and Schneider, 2001), learning biped walk patterns (Tedrake, 2004; Tedrake et al, 2005), driving a radio-controlled (RC) car, learning a jumping behavior for a robot dog (Kolter and Ng, 2009a) (illustrated in Figure 2.5), and balancing a two wheeled robot (Vlassis et al, 2009). Operational space control was also learned by Peters and Schaal (2008a) using locally linear controller models. In a marble maze task, Zucker and Bagnell (2012) used such a controller as a policy that expressed the desired velocity of the ball in terms of the directional gradient of a value function.

Non-parametric Policies. Polices based on non-parametric regression approaches often allow a more data-driven learning process. This approach is often preferable over the purely parametric policies listed above because the policy structure can evolve during the learning process. Such approaches are especially useful when a policy learned to adjust the existing behaviors of an lower-level controller, such as when choosing among different robot human interaction possibilities (Mitsunaga et al, 2005), selecting among different striking movements in a table tennis task (Peters et al, 2010a), and setting the meta-actions for dart throwing and table tennis hitting tasks (Kober et al, 2010b).

2.5 Tractability through Prior Knowledge

Prior knowledge can dramatically help guide the learning process. It can be included in the form of initial policies, demonstrations, initial models, a predefined task structure, or constraints on the policy such as torque limits or ordering constraints of the policy parameters. These approaches significantly reduce the search space and, thus, speed up the learning process. Providing a (partially) successful initial policy allows a reinforcement learning method to focus on promising regions in the value function or in policy space, see Section 2.5.1. Pre-structuring a complex task such that it can be broken down into several more tractable ones can significantly reduce the complexity of the learning task, see Section 2.5.2. An overview of publications using prior knowledge to render the learning problem tractable is presented in Table 2.4. Constraints may also limit the search space, but often pose new, additional problems for the learning methods. For example, policy search limits often do not handle hard limits on the policy well. Relaxing such constraints (a trick often applied in machine learning) is not feasible if they were introduced to protect the robot in the first place.

2.5.1 Prior Knowledge through Demonstration

People and other animals frequently learn using a combination of imitation and trial and error. When learning to play tennis, for instance, an instructor will repeatedly demonstrate the sequence of motions that form an orthodox forehand stroke. Students subsequently imitate this behavior, but still need hours of practice to successfully return balls to a precise location on the opponent's court. Input from a teacher need not be limited to initial instruction. The instructor may provide additional demonstrations in later learning stages (Latzke et al, 2007; Ross et al, 2011a) and which can also be used as differential feedback (Argall et al, 2008).

This combination of imitation learning with reinforcement learning is sometimes termed *apprenticeship learning* (Abbeel and Ng, 2004) to emphasize the need for learning both from a teacher and by practice. The term "apprenticeship learning" is often employed to refer to "inverse reinforcement learning" or "inverse optimal control" but is intended here to be employed in this original, broader meaning. For a recent survey detailing the state of the art in imitation learning for robotics, see (Argall et al, 2009).

Using demonstrations to initialize reinforcement learning provides multiple benefits. Perhaps the most obvious benefit is that it provides supervised training data of what actions to perform in states that are encountered. Such data may be helpful when used to bias policy action selection.

The most dramatic benefit, however, is that demonstration – or a hand-crafted initial policy – removes the need for *global exploration* of the policy or state-space of the RL problem. The student can improve by locally optimizing a policy knowing what states are important, making local optimization methods feasible. Intuitively, we expect that removing the demands of global exploration makes learning easier. However, we can only find local optima close to the demonstration, that is, we rely on the demonstration to provide a good starting point. Perhaps the textbook example of such in human learning is the rise of the "Fosbury Flop" (Wikipedia, 2013) method of high-jump (see Figure 2.6). This motion is very different from a classical high-jump and took generations of Olympians to discover. But after it was first demonstrated, it was soon mastered by virtually all athletes participating in the sport. On the other hand, this example also illustrates nicely that such local optimization around an initial demonstration can only find local optima.

In practice, both approximate value function based approaches and policy search methods work best for real system applications when they are constrained to make modest changes to the distribution over states while learning. Policy search approaches implicitly maintain the state distribution by limiting the changes to the policy. On the other hand, for value function methods, an unstable estimate of the value function can lead to drastic changes in the policy. Multiple policy search methods used in robotics are based on this intuition (Bagnell and Schneider, 2003; Peters and Schaal, 2008b; Peters et al, 2010a; Kober and Peters, 2011b).

Fig. 2.6 This figure illustrates the "Fosbury Flop" (public domain picture from Wikimedia Commons)

The intuitive idea and empirical evidence that demonstration makes the reinforcement learning problem simpler can be understood rigorously. In fact, Kakade and Langford (2002); Bagnell et al (2003) demonstrate that knowing approximately the state-distribution of a good policy[5] transforms the problem of reinforcement learning from one that is provably intractable in both information and computational complexity to a tractable one with only polynomial sample and computational complexity, even under function approximation and partial observability. This type of approach can be understood as a reduction from reinforcement learning to supervised learning. Both algorithms are policy search variants of approximate policy iteration that constrain policy updates. Kollar and Roy (2008) demonstrate the benefit of this RL approach for developing state-of-the-art map exploration policies and Kolter et al, 2008 employed a space-indexed variant to learn trajectory following tasks with an autonomous vehicle and a RC car.

Demonstrations by a Teacher. Demonstrations by a teacher can be obtained in two different scenarios. In the first, the teacher demonstrates the task using his or her own body; in the second, the teacher controls the robot to do the task. The first scenario is limited by the embodiment issue, as the movement of a human teacher usually cannot be mapped directly to the robot due to different physical constraints and capabilities. For example, joint angles of a human demonstrator need to be adapted to account for the kinematic differences between the teacher and the robot. Often it is more advisable to only consider task-relevant information, such asthe Cartesian positions and velocities of the end-effector and the object. Demonstrations obtained by motion-capture have been used to learn a pendulum swingup (Atkeson and Schaal, 1997), ball-in-a-cup (Kober et al, 2008) and grasping (Gräve et al, 2010).

The second scenario obtains demonstrations by a human teacher directly controlling the robot. Here the human teacher first has to learn how to achieve a task with the particular robot's hardware, adding valuable prior knowledge. For example, remotely controlling the robot initialized a Q-table for a navigation task (Conn and Peters II, 2007). If the robot is back-drivable, kinesthetic

---

[5] That is, a probability distribution over states that will be encountered when following a good policy.

teach-in (i.e., by taking it by the hand and moving it) can be employed, which enables the teacher to interact more closely with the robot. This method has resulted in applications including T-ball batting (Peters and Schaal, 2008c,b), reaching tasks (Guenter et al, 2007; Bitzer et al, 2010), ball-in-a-cup (Kober and Peters, 2008), flipping a light switch (Buchli et al, 2011), playing pool and manipulating a box (Pastor et al, 2011), and opening a door and picking up objects (Kalakrishnan et al, 2011). A marble maze task can be learned using demonstrations by a human player (Bentivegna et al, 2004a).

One of the more stunning demonstrations of the benefit of learning from a teacher is the helicopter airshows of (Coates et al, 2009). This approach combines initial human demonstration of trajectories, machine learning to extract approximate models from multiple trajectories, and classical locally-optimal control methods (Jacobson and Mayne, 1970) to achieve state-of-the-art acrobatic flight.

Hand-Crafted Policies. When human interaction with the system is not straightforward due to technical reasons or human limitations, a pre-programmed policy can provide alternative demonstrations. For example, a vision-based mobile robot docking task can be learned faster with such a basic behavior than using Q-learning alone, as demonstrated in (Martínez-Marín and Duckett, 2005) Providing hand-coded, stable initial gaits can significantly help in learning robot locomotion, as shown on a six-legged robot (Erden and Leblebicioğlu, 2008) as well as on a biped (Tedrake, 2004; Tedrake et al, 2005). Alternatively, hand-crafted policies can yield important corrective actions as prior knowledge that prevent the robot to deviates significantly from the desired behavior and endanger itself. This approach has been applied to adapt the walking patterns of a robot dog to new surfaces (Birdwell and Livingston, 2007) by Q-learning. Rosenstein and Barto (2004) employed a stable controller to teach the robot about favorable actions and avoid risky behavior while learning to move from a start to a goal position.

2.5.2 *Prior Knowledge through Task Structuring*

Often a task can be decomposed hierarchically into basic components or into a sequence of increasingly difficult tasks. In both cases the complexity of the learning task is significantly reduced.

Hierarchical Reinforcement Learning. A task can often be decomposed into different levels. For example when using meta-actions (Section 2.4.1), these meta-actions correspond to a lower level dealing with the execution of sub-tasks which are coordinated by a strategy level. Hierarchical reinforcement learning does not assume that all but one levels are fixed but rather learns all of them simultaneously. For example, hierarchical Q-learning has been used to learn different behavioral levels for a six legged robot: moving single legs, locally moving the complete body, and globally moving the robot towards a

Table 2.4 This table illustrates different methods of making robot reinforcement learning tractable by incorporating prior knowledge

PRIOR KNOWLEDGE THROUGH DEMONSTRATION

Approach	Employed by...
Teacher	Atkeson and Schaal (1997); Bentivegna et al (2004a); Bitzer et al (2010); Conn and Peters II (2007); Gräve et al (2010); Kober et al (2008); Kober and Peters (2008); Latzke et al (2007); Peters and Schaal (2008c,b)
Policy	Birdwell and Livingston (2007); Erden and Leblebicioğlu (2008); Martínez-Marín and Duckett (2005); Rosenstein and Barto (2004); Smart and Kaelbling (1998); Tedrake (2004); Tedrake et al (2005); Wang et al (2006)

PRIOR KNOWLEDGE THROUGH TASK STRUCTURING

Approach	Employed by...
Hierarchical	Daniel et al (2012); Donnart and Meyer (1996); Hart and Grupen (2011); Huber and Grupen (1997); Kirchner (1997); Morimoto and Doya (2001); Muelling et al (2012); Whitman and Atkeson (2010)
Progressive Tasks	Asada et al (1996); Randløv and Alstrøm (1998)

DIRECTED EXPLORATION WITH PRIOR KNOWLEDGE

Approach	Employed by...
Directed Exploration	Huang and Weng (2002); Kroemer et al (2010); Pendrith (1999)

goal (Kirchner, 1997). A stand-up behavior considered as a hierarchical reinforcement learning task has been learned using Q-learning in the upper-level and a continuous actor-critic method in the lower level (Morimoto and Doya, 2001). Navigation in a maze can be learned using an actor-critic architecture by tuning the influence of different control modules and learning these modules (Donnart and Meyer, 1996). Huber and Grupen (1997) combine discrete event system and reinforcement learning techniques to learn turning gates for a quadruped. Hart and Grupen (2011) learn to bi-manual manipulation tasks by assembling policies hierarchically. Daniel et al (2012) learn options in a tetherball scenario and Muelling et al (2012) learn different strokes in a table tennis scenario. Whitman and Atkeson (2010) show that the optimal policy for some global systems (like a walking controller) can be constructed by finding the optimal controllers for simpler subsystems and coordinating these.

Progressive Tasks. Often complicated tasks are easier to learn if simpler tasks can already be performed. This progressive task development is inspired by how biological systems learn. For example, a baby first learns how to roll, then how to crawl, then how to walk. A sequence of increasingly difficult missions has been employed to learn a goal shooting task in (Asada et al, 1996) using Q-learning. Randløv and Alstrøm (1998) discuss shaping the reward function to include both a balancing and a goal oriented term for a simulated bicycle riding task. The reward is constructed in such a way that the balancing term dominates the other term and, hence, this more fundamental behavior is learned first.

2.5.3   Directing Exploration with Prior Knowledge

As discussed in Section 2.2.1, balancing exploration and exploitation is an important consideration. Task knowledge can be employed to guide to robots curiosity to focus on regions that are novel and promising at the same time. For example, a mobile robot learns to direct attention by employing a modified Q-learning approach using novelty (Huang and Weng, 2002). Using "corrected truncated returns" and taking into account the estimator variance, a six legged robot employed with stepping reflexes can learn to walk (Pendrith, 1999). Offline search can be used to guide Q-learning during a grasping task (Wang et al, 2006). Using upper confidence bounds (Kaelbling, 1990) to direct exploration into regions with potentially high rewards, grasping can be learned efficiently (Kroemer et al, 2010).

2.6   Tractability through Models

In Section 2.2, we discussed robot reinforcement learning from a model-free perspective where the system simply served as a data generating process. Such model-free reinforcement algorithms try to directly learn the value function or the policy without any explicit modeling of the transition dynamics. In contrast, many robot reinforcement learning problems can be made tractable by learning forward models, i.e., approximations of the transition dynamics based on data. Such model-based reinforcement learning approaches jointly learn a model of the system with the value function or the policy and often allow for training with less interaction with the the real environment. Reduced learning on the real robot is highly desirable as simulations are frequently faster than real-time while safer for both the robot and its environment. The idea of combining learning in simulation and in the real environment was popularized by the Dyna-architecture (Sutton, 1990), prioritized sweeping (Moore and Atkeson, 1993), and incremental multi-step Q-learning (Peng and Williams, 1996) in reinforcement learning. In robot reinforcement learning, the learning step on the simulated system is often called "mental rehearsal". We first discuss the core issues and techniques in mental rehearsal for robotics (Section 2.6.1), and, subsequently, we discuss learning methods that have be used in conjunction with learning with forward models (Section 2.6.2). An overview of publications using simulations to render the learning problem tractable is presented in Table 2.5.

2.6.1 Core Issues and General Techniques in Mental Rehearsal

Experience collected in the real world can be used to learn a forward model (Åström and Wittenmark, 1989) from data. Such forward models allow training by interacting with a simulated environment. Only the resulting policy is subsequently transferred to the real environment. Model-based methods can make the learning process substantially more sample efficient. However, depending on the type of model, these methods may require a great deal of memory. In the following paragraphs, we deal with the core issues of mental rehearsal: simulation biases, stochasticity of the real world, and efficient optimization when sampling from a simulator.

Dealing with Simulation Biases. It is impossible to obtain a forward model that is accurate enough to simulate a complex real-world robot system without error. If the learning methods require predicting the future or using derivatives, even small inaccuracies can quickly accumulate, significantly amplifying noise and errors (An et al, 1988). Reinforcement learning approaches exploit such model inaccuracies if they are beneficial for the reward received in simulation (Atkeson and Schaal, 1997). The resulting policies may work well with the forward model (i.e., the simulator) but poorly on the real system. This is known as simulation bias. It is analogous to over-fitting in supervised learning – that is, the algorithm is doing its job well on the model and the training data, respectively, but does not generalize well to the real system or novel data. Simulation bias often leads to biased, potentially physically non-feasible solutions while even iterating between model learning and policy will have slow convergence. Averaging over the model uncertainty in probabilistic models can be used to reduce the bias; see the next paragraph for examples. Another result from these simulation biases is that relatively few researchers have successfully demonstrated that a policy learned in simulation can directly be transferred to a real robot while maintaining a high level of performance. The few examples include maze navigation tasks (Bakker et al, 2003; Oßwald et al, 2010; Youssef, 2005), obstacle avoidance (Fagg et al, 1998) for a mobile robot, very basic robot soccer (Duan et al, 2007) and multi-legged robot locomotion (Ilg et al, 1999; Svinin et al, 2001). Nevertheless, simulation biases can be addressed by introducing stochastic models or distributions over models even if the system is very close to deterministic. Artificially adding a little noise will smooth model errors and avoid policy over-fitting (Jakobi et al, 1995; Atkeson, 1998). On the downside, potentially very precise policies may be eliminated due to their fragility in the presence of noise. This technique can be beneficial in all of the approaches described in this section. Nevertheless, in recent work, Ross and Bagnell (2012) presented an approach with strong guarantees for learning the model and policy in an iterative fashion even if the true system is not in the model class, indicating that it may be possible to deal with simulation bias.

Fig. 2.7 Autonomous inverted helicopter flight (Ng et al, 2004b)(Picture reprint with permission of Andrew Ng)

Distributions over Models for Simulation. Model learning methods that maintain probabilistic uncertainty about true system dynamics allow the RL algorithm to generate distributions over the performance of a policy. Such methods explicitly model the uncertainty associated with the dynamics at each state and action. For example, when using a Gaussian process model of the transition dynamics, a policy can be evaluated by propagating the state and associated uncertainty forward in time. Such evaluations in the model can be used by a policy search approach to identify where to collect more data to improve a policy, and may be exploited to ensure that control is safe and robust to model uncertainty (Schneider, 1996; Bagnell and Schneider, 2001). When the new policy is evaluated on the real system, the novel observations can subsequently be incorporated into the forward model. Bagnell and Schneider (2001) showed that maintaining model uncertainty and using it in the inner-loop of a policy search method enabled effective flight control using only minutes of collected data, while performance was compromised by considering a best-fit model. This approach uses explicit Monte-Carlo simulation in the sample estimates.

By treating model uncertainty as if it were noise (Schneider, 1996) as well as employing analytic approximations of forward simulation, a cart-pole task can be solved with less than 20 seconds of interaction with the physical system (Deisenroth and Rasmussen, 2011); a visually driven block-stacking task has also been learned data-efficiently (Deisenroth et al, 2011). Similarly, solving a linearized control problem with multiple probabilistic models and combining the resulting closed-loop control with open-loop control has resulted in autonomous sideways sliding into a parking spot (Kolter et al, 2010). Instead of learning a model of the system dynamics, Lizotte et al (2007) directly learned the expected return as a function of the policy parameters using Gaussian process regression in a black-box fashion, and, subsequently, searched for promising parameters in this model. The method has been applied to optimize the gait of an Aibo robot.

Sampling by Re-using Random Numbers. A forward model can be used as a simulator to create roll-outs for training by sampling. When comparing results of different simulation runs, it is often hard to tell from a small number

Table 2.5 This table illustrates different methods of making robot reinforcement learning tractable using models

CORE ISSUES AND GENERAL TECHNIQUES IN MENTAL REHEARSAL

Approach	Employed by...
Dealing with Simulation Biases	An et al (1988); Atkeson and Schaal (1997); Atkeson (1998); Bakker et al (2003); Duan et al (2007); Fagg et al (1998); Ilg et al (1999); Jakobi et al (1995); Oßwald et al (2010); Ross and Bagnell (2012); Svinin et al (2001); Youssef (2005)
Distributions over Models for Simulation	Bagnell and Schneider (2001); Deisenroth and Rasmussen (2011); Deisenroth et al (2011); Kolter et al (2010); Lizotte et al (2007)
Sampling by Re-Using Random Numbers	Bagnell and Schneider (2001); Bagnell (2004); Ko et al (2007); Michels et al (2005); Ng et al (2004a,b)

SUCCESSFUL LEARNING APPROACHES WITH FORWARD MODELS

Approach	Employed by...
Iterative Learning Control	Abbeel et al (2006); An et al (1988); Berg et al (2010); Bukkems et al (2005); Freeman et al (2010); Norrlöf (2002)
Locally Linear Quadratic Regulators	Atkeson and Schaal (1997); Atkeson (1998); Coates et al (2009); Kolter et al (2008); Schaal and Atkeson (1994); Tedrake et al (2010)
Value Function Methods with Learned Models	Bakker et al (2006); Nemec et al (2010); Uchibe et al (1998)
Policy Search with Learned Models	Bagnell and Schneider (2001); Bagnell (2004); Deisenroth et al (2011); Deisenroth and Rasmussen (2011); Kober and Peters (2011b); Ng et al (2004a,b); Peters et al (2010a)

of samples whether a policy really worked better or whether the results are an effect of the simulated stochasticity. Using a large number of samples to obtain proper estimates of the expectations become prohibitively expensive if a large number of such comparisons need to be performed (e.g., for gradient estimation within an algorithm). A common technique in the statistics and simulation community (Glynn, 1987) to address this problem is to re-use the series of random numbers in fixed models, hence mitigating the noise contribution. Ng et al (2004a,b) extended this approach for learned simulators. The resulting approach, PEGASUS, found various applications in the learning of maneuvers for autonomous helicopters (Bagnell and Schneider, 2001; Bagnell, 2004; Ng et al, 2004a,b), as illustrated in Figure 2.7. It has been used to learn control parameters for a RC car (Michels et al, 2005) and an autonomous blimp (Ko et al, 2007).

While mental rehearsal has a long history in robotics, it is currently becoming again a hot topic, especially due to the work on probabilistic virtual simulation.

2.6.2   Successful Learning Approaches with Forward Models

Model-based approaches rely on an approach that finds good policies using the learned model. In this section, we discuss methods that directly obtain a new policy candidate directly from a forward model. Some of these methods have a long history in optimal control and only work in conjunction with a forward model.

Iterative Learning Control. A powerful idea that has been developed in multiple forms in both the reinforcement learning and control communities is the use of crude, approximate models to determine gradients, e.g., for an update step. The resulting new policy is then evaluated in the real world and the model is updated. This approach is known as iterative learning control (Arimoto et al, 1984). A similar preceding idea was employed to minimize trajectory tracking errors (An et al, 1988) and is loosely related to feedback error learning (Kawato, 1990). More recently, variations on the iterative learning control has been employed to learn robot control (Norrlöf, 2002; Bukkems et al, 2005), steering a RC car with a general analysis of approximate models for policy search in (Abbeel et al, 2006), a pick and place task (Freeman et al, 2010), and an impressive application of tying knots with a surgical robot at superhuman speeds (Berg et al, 2010).

Locally Linear Quadratic Regulators. Instead of sampling from a forward model-based simulator, such learned models can be directly used to compute optimal control policies. This approach has resulted in a variety of robot reinforcement learning applications that include pendulum swing-up tasks learned with DDP (Atkeson and Schaal, 1997; Atkeson, 1998), devil-sticking (a form of gyroscopic juggling) obtained with local LQR solutions (Schaal and Atkeson, 1994), trajectory following with space-indexed controllers trained with DDP for an autonomous RC car (Kolter et al, 2008), and the aerobatic helicopter flight trained with DDP discussed above (Coates et al, 2009).

Value Function Methods with Learned Models. Obviously, mental rehearsal can be used straightforwardly with value function methods by simply pretending that the simulated roll-outs were generated from the real system. Learning in simulation while the computer is idle and employing directed exploration allows Q-learning to learn a navigation task from scratch in 20 minutes (Bakker et al, 2006). Two robots taking turns in learning a simplified soccer task were also able to profit from mental rehearsal (Uchibe et al, 1998). Nemec et al (2010) used a value function learned in simulation to initialize the real robot learning. However, it is clear that model-based methods that use the model for creating more direct experience should potentially perform better.

Policy Search with Learned Models. Similarly as for value function methods, all model-free policy search methods can be used in conjunction with learned simulators. For example, both pairwise comparisons of policies and policy gradient approaches have been used with learned simulators (Bagnell and Schneider, 2001; Bagnell, 2004; Ng et al, 2004a,b). Transferring EM-like policy search Kober and Peters (2011b) and Relative Entropy Policy Search Peters et al (2010a) appears to be a natural next step. Nevertheless, as mentioned in Section 2.6.1, a series of policy update methods has been suggested that were tailored for probabilistic simulation Deisenroth et al (2011); Deisenroth and Rasmussen (2011); Lizotte et al (2007).

There still appears to be the need for new methods that make better use of the model knowledge, both in policy search and for value function methods.

2.7 A Case Study: Ball-in-a-Cup

Up to this point in this chapter, we have reviewed a large variety of problems and associated solutions within robot reinforcement learning. In this section, we will take a complementary approach and discuss one task in detail that has previously been studied.

This *ball-in-a-cup* task due to its relative simplicity can serve as an example to highlight some of the challenges and methods that were discussed earlier. We do not claim that the method presented is the best or only way to address the presented problem; instead, our goal is to provide a case study that shows design decisions which can lead to successful robotic reinforcement learning.

In Section 2.7.1, the experimental setting is described with a focus on the task and the reward. Section 2.7.2 discusses a type of pre-structured policies that has been particularly useful in robotics. Inclusion of prior knowledge is presented in Section 2.7.3. The advantages of the employed policy search algorithm are explained in Section 2.7.4. The use of simulations in this task is discussed in Section 2.7.5 and results on the real robot are described in Section 2.7.6. Finally, an alternative reinforcement learning approach is briefly explored in Section 2.7.7.

2.7.1 Experimental Setting: Task and Reward

The children's game ball-in-a-cup, also known as *balero* and *bilboquet*, is challenging even for adults. The toy consists of a small cup held in one hand (in this case, it is attached to the end-effector of the robot) and a small ball hanging on a string attached to the cup's bottom (for the employed toy, the string is 40cm long). Initially, the ball is at rest, hanging down vertically. The player needs to move quickly to induce motion in the ball through the string, toss the ball in the air, and catch it with the cup. A possible movement is illustrated in Figure 2.8a.

The state of the system can be described by joint angles and joint velocities of the robot as well as the Cartesian coordinates and velocities of the ball (neglecting states that cannot be observed straightforwardly like the state of

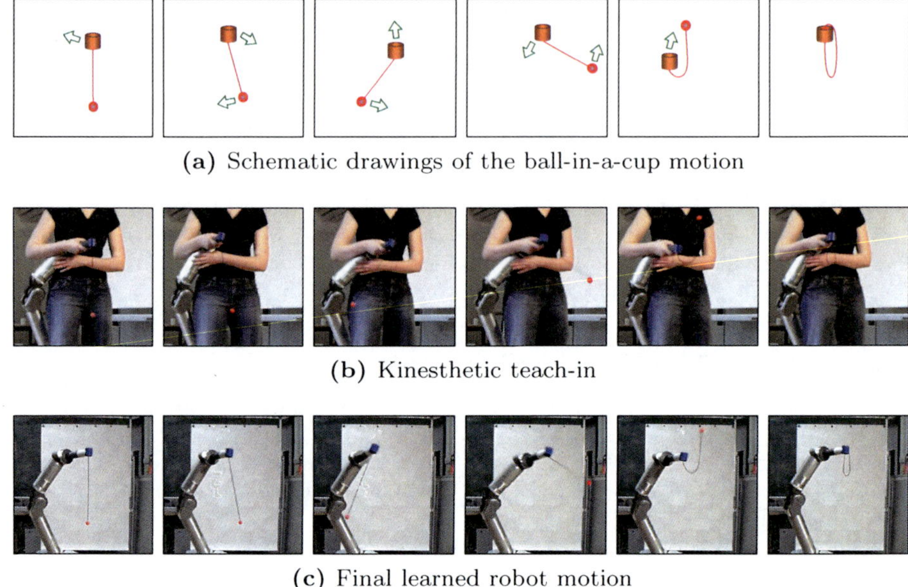

(a) Schematic drawings of the ball-in-a-cup motion

(b) Kinesthetic teach-in

(c) Final learned robot motion

Fig. 2.8 This figure shows schematic drawings of the ball-in-a-cup motion (a), the final learned robot motion (c), as well as a kinesthetic teach-in (b). The green arrows show the directions of the current movements in that frame. The human cup motion was taught to the robot by imitation learning. The robot manages to reproduce the imitated motion quite accurately, but the ball misses the cup by several centimeters. After approximately 75 iterations of the Policy learning by Weighting Exploration with the Returns (PoWER) algorithm the robot has improved its motion so that the ball regularly goes into the cup.

the string or global room air movement). The actions are the joint space accelerations, which are translated into torques by a fixed inverse dynamics controller. Thus, the reinforcement learning approach has to deal with twenty state and seven action dimensions, making discretization infeasible.

An obvious reward function would be a binary return for the whole episode, depending on whether the ball was caught in the cup or not. In order to give the reinforcement learning algorithm a notion of closeness, Kober and Peters (2011b) initially used a reward function based solely on the minimal distance between the ball and the cup. However, the algorithm has exploited rewards resulting from hitting the cup with the ball from below or from the side, as such behaviors are easier to achieve and yield comparatively high rewards. To avoid such local optima, it was essential to find a good reward function that contains the additional prior knowledge that getting the ball into the cup is only possible from one direction. Kober and Peters (2011b) expressed this constraint by computing the reward as $r(t_c) = \exp\left(-\alpha(x_c - x_b)^2 - \alpha(y_c - y_b)^2\right)$ while $r(t) = 0$ for all $t \neq t_c$. Here,

t_c corresponds to the time step when the ball passes the rim of the cup with a downward direction, the cup position is denoted by $[x_c, y_c, z_c] \in \mathbb{R}^3$, the ball position is $[x_b, y_b, z_b] \in \mathbb{R}^3$ and the scaling parameter $\alpha = 100$. This reward function does not capture the case where the ball does not pass above the rim of the cup, as the reward will always be zero. The employed approach performs a local policy search, and hence, an initial policy that brings the ball above the rim of the cup was required.

The task exhibits some surprising complexity as the reward is not only affected by the cup's movements but foremost by the ball's movements. As the ball's movements are very sensitive to small perturbations, the initial conditions, or small arm movement changes will drastically affect the outcome. Creating an accurate simulation is hard due to the nonlinear, unobservable dynamics of the string and its non-negligible weight.

2.7.2 Appropriate Policy Representation

The policy is represented by dynamical systems motor primitives (Ijspeert et al, 2002a; Schaal et al, 2007). The global movement is encoded as a point attractor linear dynamical system with an additional local transformation that allows a very parsimonious representation of the policy. This framework ensures the stability of the movement and allows the representation of arbitrarily shaped movements through the primitive's policy parameters. These parameters can be estimated straightforwardly by locally weighted regression. Note that the dynamical systems motor primitives ensure the stability of the movement generation but cannot guarantee the stability of the movement execution. These primitives can be modified through their meta-parameters in order to adapt to the final goal position, the movement amplitude, or the duration of the movement. The resulting movement can start from arbitrary positions and velocities and go to arbitrary final positions while maintaining the overall shape of the trajectory. See Chapter 3 for a more detailed discussion.

This policy can be seen as a parameterization of a mean policy in the form $\bar{a} = \boldsymbol{\theta}^{\mathrm{T}} \boldsymbol{\mu}(\mathbf{s}, t)$, which is linear in parameters. Thus, it is straightforward to include prior knowledge from a demonstration using supervised learning by locally weighted regression.

This policy is augmented by an additive exploration $\boldsymbol{\epsilon}(\mathbf{s}, t)$ noise term to make policy search methods possible. As a result, the explorative policy can be given in the form $\mathbf{a} = \boldsymbol{\theta}^{\mathrm{T}} \boldsymbol{\mu}(\mathbf{s}, t) + \boldsymbol{\epsilon}(\boldsymbol{\mu}(\mathbf{s}, t))$. Some policy search approaches have previously used state-independent, white Gaussian exploration, i.e., $\boldsymbol{\epsilon}(\boldsymbol{\mu}(\mathbf{s}, t)) \sim \mathcal{N}(0, \boldsymbol{\Sigma})$. However, such unstructured exploration at every step has several disadvantages, notably: (i) it causes a large variance which grows with the number of time-steps, (ii) it perturbs actions too frequently, thus, "washing out" their effects and (iii) it can damage the system that is executing the trajectory.

Alternatively, one could generate a form of structured, state-dependent exploration (Rückstieß et al, 2008) $\epsilon(\boldsymbol{\mu}(\mathbf{s},t)) = \boldsymbol{\varepsilon}_t^{\mathrm{T}} \boldsymbol{\mu}(\mathbf{s},t)$ with $[\varepsilon_t]_{ij} \sim \mathcal{N}(0,\sigma_{ij}^2)$, where σ_{ij}^2 are meta-parameters of the exploration that can also be optimized. This argument results in the policy $\mathbf{a} = (\boldsymbol{\theta} + \boldsymbol{\varepsilon}_t)^{\mathrm{T}} \boldsymbol{\mu}(\mathbf{s},t)$ corresponding to the distribution $\mathbf{a} \sim \pi(\mathbf{a}_t|\mathbf{s}_t,t) = \mathcal{N}(\mathbf{a}|\boldsymbol{\theta}^{\mathrm{T}}\boldsymbol{\mu}(\mathbf{s},t), \boldsymbol{\mu}(\mathbf{s},t)^{\mathrm{T}}\hat{\boldsymbol{\Sigma}}(\mathbf{s},t))$. Instead of directly exploring in action space, this type of policy explores in parameter space.

2.7.3   Generating a Teacher Demonstration

Children usually learn this task by first observing another person presenting a demonstration. They then try to duplicate the task through trial-and-error-based learning. To mimic this process, the motor primitives were first initialized by imitation. Subsequently, they were improved them by reinforcement learning.

A demonstration for imitation was obtained by recording the motions of a human player performing kinesthetic teach-in as shown in Figure 2.8b. Kinesthetic teach-in means "taking the robot by the hand", performing the task by moving the robot while it is in gravity-compensation mode, and recording the joint angles, velocities and accelerations. It requires a back-drivable robot system that is similar enough to a human arm to not cause embodiment issues. Even with demonstration, the resulting robot policy fails to catch the ball with the cup, leading to the need for self-improvement by reinforcement learning. As discussed in Section 2.7.1, the initial demonstration was needed to ensure that the ball goes above the rim of the cup.

2.7.4   Reinforcement Learning by Policy Search

Policy search methods are better suited for a scenario like this, where the task is episodic, local optimization is sufficient (thanks to the initial demonstration), and high dimensional, continuous states and actions need to be taken into account. A single update step in a gradient based method usually requires as many episodes as parameters to be optimized. Since the expected number of parameters was in the hundreds, a different approach had to be taken because gradient based methods are impractical in this scenario. Furthermore, the step-size parameter for gradient based methods often is a crucial parameter that needs to be tuned to achieve good performance. Instead, an expectation-maximization inspired algorithm was employed that requires significantly less samples and has no learning rate.

Kober and Peters (2008) have derived a framework of reward weighted imitation. See Chapter 4 for a more detailed discussion. Based on (Dayan and Hinton, 1997) they consider the return of an episode as an improper probability distribution. A lower bound of the logarithm of the expected return is maximized. Depending on the strategy of optimizing this lower bound and the exploration strategy, the framework yields several well known policy search algorithms as well as the novel Policy learning by Weighting Exploration with the Returns (PoWER) algorithm. PoWER is an expectation-maximization inspired algorithm that employs state-dependent exploration (as discussed in Section 2.7.2). The update rule is given by

$$\boldsymbol{\theta}' = \boldsymbol{\theta} + E\left\{\sum_{t=1}^{T} \mathbf{W}\left(\mathbf{s}_t, t\right) Q^{\pi}\left(\mathbf{s}_t, \mathbf{a}_t, t\right)\right\}^{-1}$$
$$E\left\{\sum_{t=1}^{T} \mathbf{W}\left(\mathbf{s}_t, t\right) \boldsymbol{\varepsilon}_t Q^{\pi}\left(\mathbf{s}_t, \mathbf{a}_t, t\right)\right\},$$

where $\mathbf{W}\left(\mathbf{s}_t, t\right) = \boldsymbol{\mu}(\mathbf{s}, t)\boldsymbol{\mu}(\mathbf{s}, t)^{\mathrm{T}}(\boldsymbol{\mu}(\mathbf{s}, t)^{\mathrm{T}}\hat{\boldsymbol{\Sigma}}\boldsymbol{\mu}(\mathbf{s}, t))^{-1}$. Intuitively, this update can be seen as a reward-weighted imitation, (or recombination) of previously seen episodes. Depending on its effect on the state-action value function, the exploration of each episode is incorporated more or less strongly into the updated policy. To reduce the number of trials in this on-policy scenario, the trials are reused through importance sampling (Sutton and Barto, 1998). To avoid the fragility that sometimes results from importance sampling in reinforcement learning, samples with very small importance weights were discarded. In essence, this algorithm performs a local search around the policy learned from demonstration and prior knowledge.

2.7.5   Use of Simulations in Robot Reinforcement Learning

The robot is simulated by rigid body dynamics with parameters estimated from data. The toy is simulated as a pendulum with an elastic string that switches to a ballistic point mass when the ball is closer to the cup than the string is long. The spring, damper and restitution constants were tuned to match data recorded on a VICON system. The SL framework (Schaal, 2009) allowed us to switch between the simulated robot and the real one with a simple recompile. Even though the simulation matches recorded data very well, policies that get the ball in the cup in simulation usually missed the cup by several centimeters on the real system and vice-versa. One conceivable approach could be to first improve a demonstrated policy in simulation and only perform the fine-tuning on the real robot.

However, this simulation was very helpful to develop and tune the algorithm as it runs faster in simulation than real-time and does not require human supervision or intervention. The algorithm was initially confirmed and tweaked with unrelated, simulated benchmark tasks (shown in (Kober and Peters, 2011b)). The use of an importance sampler was essential to achieve good performance and required many tests over the course of two weeks. A very crude importance sampler that considers only the n best previous episodes worked sufficiently well in practice. Depending on the number n the algorithm exhibits are more or less pronounced greedy behavior. Additionally there are a number of possible simplifications for the learning algorithm, some of which work very well in practice even if the underlying assumption do not strictly hold in reality. The finally employed variant

$$\boldsymbol{\theta}' = \boldsymbol{\theta} + \frac{\left\langle \sum_{t=1}^{T} \varepsilon_t Q^{\pi}(\mathbf{s}_t, \mathbf{a}_t, t) \right\rangle_{\omega(\tau)}}{\left\langle \sum_{t=1}^{T} Q^{\pi}(\mathbf{s}_t, \mathbf{a}_t, t) \right\rangle_{\omega(\tau)}}$$

assumes that only a single basis function is active at a given time, while there is actually some overlap for the motor primitive basis functions. The importance sampler is denoted by $\langle \cdot \rangle_{\omega(\tau)}$. The implementation is further simplified as the reward is zero for all but one time-step per episode.

To adapt the algorithm to this particular task, the most important parameters to tune were the "greediness" of the importance sampling, the initial magnitude of the exploration, and the number of parameters for the policy. These parameters were identified by a coarse grid search in simulation with various initial demonstrations and seeds for the random number generator. Once the simulation and the grid search were coded, this process only took a few minutes. The exploration parameter is fairly robust if it is in the same order of magnitude as the policy parameters. For the importance sampler, using the 10 best previous episodes was a good compromise. The number of policy parameters needs to be high enough to capture enough details to get the ball above the rim of the cup for the initial demonstration. On the other hand, having more policy parameters will potentially slow down the learning process. The number of needed policy parameters for various demonstrations were in the order of 30 parameters per DoF. The demonstration employed for the results shown in more detail in this chapter employed 31 parameters per DoF for an approximately 3 second long movement, hence 217 policy parameters total. Having three times as many policy parameters slowed down the convergence only slightly.

2.7.6 Results on the Real Robot

The first run on the real robot used the demonstration shown in Figure 2.8 and directly worked without any further parameter tuning. For the five runs with this demonstration, which took approximately one hour each, the robot got the ball into the cup for the first time after 42-45 episodes and regularly succeeded at bringing the ball into the cup after 70-80 episodes. The policy always converged to the maximum after 100 episodes. Running the real robot experiment was tedious as the ball was tracked by a stereo vision system, which sometimes failed and required a manual correction of the reward. As the string frequently entangles during failures and the robot cannot unravel it, human intervention is required. Hence, the ball had to be manually reset after each episode.

If the constraint of getting the ball above the rim of the cup for the initial policy is fulfilled, the presented approach works well for a wide variety of initial demonstrations including various teachers and two different movement strategies (swinging the ball or pulling the ball straight up). Convergence took between 60 and 130 episodes, which largely depends on the initial distance to the cup but also on the robustness of the demonstrated policy.

2.7.7 *Alternative Approach with Value Function Methods*

Nemec et al (2010) employ an alternate reinforcement learning approach to achieve the ball-in-a-cup task with a Mitsubishi PA10 robot. They decomposed the task into two sub-tasks, the swing-up phase and the catching phase. In the swing-up phase, the ball is moved above the cup. In the catching phase, the ball is caught with the cup using an analytic prediction of the ball trajectory based on the movement of a flying point mass. The catching behavior is fixed; only the swing-up behavior is learned. The paper proposes to use SARSA to learn the swing-up movement. The states consist of the cup positions and velocities as well as the angular positions and velocities of the ball. The actions are the accelerations of the cup in a single Cartesian direction. Tractability is achieved by discretizing both the states (324 values) and the actions (5 values) and initialization by simulation. The behavior was first learned in simulation requiring 220 to 300 episodes. The state-action value function learned in simulation was used to initialize the learning on the real robot. The robot required an additional 40 to 90 episodes to adapt the behavior learned in simulation to the real environment.

2.8 Discussion

We have surveyed the state of the art in robot reinforcement learning for both general reinforcement learning audiences and robotics researchers to provide possibly valuable insight into successful techniques and approaches.

From this overview, it is clear that using reinforcement learning in the domain of robotics is not yet a straightforward undertaking but rather requires a certain amount of skill. Hence, in this section, we highlight several open questions faced by the robotic reinforcement learning community in order to make progress towards "off-the-shelf" approaches as well as a few current practical challenges. Finally, we try to summarize several key lessons from robotic reinforcement learning for the general reinforcement learning community.

2.8.1 Open Questions in Robotic Reinforcement Learning

Reinforcement learning is clearly not applicable to robotics "out of the box" yet, in contrast to supervised learning where considerable progress has been made in large-scale, easy deployment over the last decade. As this chapter illustrates, reinforcement learning can be employed for a wide variety of physical systems and control tasks in robotics. It is unlikely that single answers do exist for such a heterogeneous field, but even for very closely related tasks, appropriate methods currently need to be carefully selected. The user has to decide when sufficient prior knowledge is given and learning can take over. All methods require hand-tuning for choosing appropriate representations, reward functions, and the required prior knowledge. Even the correct use of models in robot reinforcement learning requires substantial future research. Clearly, a key step in robotic reinforcement learning is the automated choice of these elements and having robust algorithms that limit the required tuning to the point where even a naive user would be capable of using robotic RL.

How to choose representations automatically? The automated selection of appropriate representations remains a difficult problem as the action space in robotics often is inherently continuous and multi-dimensional. While there are good reasons for using the methods presented in Section 2.4 in their respective domains, the question whether to approximate states, value functions, policies – or a mix of the three – remains open. The correct methods to handle the inevitability of function approximation remain under intense study, as does the theory to back it up.

How to generate reward functions from data? Good reward functions are *always* essential to the success of a robot reinforcement learning approach (see Section 2.3.4). While inverse reinforcement learning can be used as an alternative to manually designing the reward function, it relies on the design of features that capture the important aspects of the problem space instead. Finding feature candidates may require insights not altogether different from the ones needed to design the actual reward function.

How much can prior knowledge help? How much is needed? Incorporating prior knowledge is one of the main tools to make robotic reinforcement learning tractable (see Section 2.5). However, it is often hard to tell in advance how much prior knowledge is required to enable a reinforcement learning algorithm to succeed in a reasonable number of episodes. For such cases, a loop of imitation and reinforcement learning may be a desirable alternative. Nevertheless, sometimes, prior knowledge may not help at all. For example, obtaining initial policies from human demonstrations can be virtually impossible if the morphology of the robot is too different from a human's (which is known as the correspondence problem (Argall et al, 2009)). Whether alternative forms of prior knowledge can help here may be a key question to answer.

How to integrate more tightly with perception? Much current work on robotic reinforcement learning relies on subsystems that abstract away perceptual information, limiting the techniques to simple perception systems and heavily pre-processed data. This abstraction is in part due to limitations of existing reinforcement learning approaches at handling inevitably incomplete, ambiguous and noisy sensor data. Learning active perception jointly with the robot's movement and semantic perception are open problems that present tremendous opportunities for simplifying as well as improving robot behavior programming.

How to reduce parameter sensitivity? Many algorithms work fantastically well for a very narrow range of conditions or even tuning parameters. A simple example would be the step size in a gradient based method. However, searching anew for the best parameters for each situation is often prohibitive with physical systems. Instead algorithms that work fairly well for a large range of situations and parameters are potentially much more interesting for practical robotic applications.

How to deal with model errors and under-modeling? Model based approaches can significantly reduce the need for real-world interactions. Methods that are based on approximate models and use local optimization often work well (see Section 2.6). As real-world samples are usually more expensive than comparatively cheap calculations, this may be a significant advantage. However, for most robot systems, there will always be under-modeling and resulting model errors. Hence, the policies learned only in simulation frequently cannot be transferred directly to the robot.

This problem may be inevitable due to both uncertainty about true system dynamics, the non-stationarity of system dynamics[6] and the inability of any model in our class to be perfectly accurate in its description of those dynamics, which have led to robust control theory (Zhou and Doyle, 1997). Reinforcement learning approaches mostly require the behavior designer to deal with this problem by incorporating model uncertainty with artificial noise or carefully choosing reward functions to discourage controllers that generate frequencies that might excite unmodeled dynamics.

Tremendous theoretical and algorithmic challenges arise from developing algorithms that are robust to both model uncertainty and under-modeling while ensuring fast learning and performance. A possible approach may be the full Bayesian treatment of the impact of possible model errors onto the policy but has the risk of generating overly conservative policies, as also happened in robust optimal control.

This list of questions is by no means exhaustive, however, it provides a fair impression of the critical issues for basic research in this area.

---

[6] Both the environment and the robot itself change dynamically; for example, vision systems depend on the lighting condition and robot dynamics change with wear and the temperature of the grease.

2.8.2 Practical Challenges for Robotic Reinforcement Learning

More applied problems for future research result from practical challenges in robot reinforcement learning:

Exploit Data Sets Better. Humans who learn a new task build upon previously learned skills. For example, after a human learns how to throw balls, learning how to throw darts becomes significantly easier. Being able to transfer previously-learned skills to other tasks and potentially to robots of a different type is crucial. For complex tasks, learning cannot be achieved globally. It is essential to reuse other locally learned information from past data sets. While such transfer learning has been studied more extensively in other parts of machine learning, tremendous opportunities for leveraging such data exist within robot reinforcement learning. Making such data sets with many skills publicly available would be a great service for robotic reinforcement learning research.

Comparable Experiments and Consistent Evaluation. The difficulty of performing, and reproducing, large scale experiments due to the expense, fragility, and differences between hardware remains a key limitation in robotic reinforcement learning. The current movement towards more standardization within robotics may aid these efforts significantly, e.g., by possibly providing a standard robotic reinforcement learning setup to a larger community – both real and simulated. These questions need to be addressed in order for research in self-improving robots to progress.

2.8.3 Robotics Lessons for Reinforcement Learning

Most of the article has aimed equally at both reinforcement learning and robotics researchers as well as practitioners. However, this section attempts to convey a few important, possibly provocative, take-home messages for the classical reinforcement learning community.

Focus on High-Dimensional Continuous Actions and Constant Adaptation. Robotic problems clearly have driven theoretical reinforcement learning research, particularly in policy search, inverse optimal control approaches, and model robustness. The practical impact of robotic reinforcement learning problems (e.g., multi-dimensional continuous action spaces, continuously drifting noise, frequent changes in the hardware and the environment, and the inevitability of undermodeling), may not yet have been sufficiently appreciated by the theoretical reinforcement learning community. These problems have often caused robotic reinforcement learning to take significantly different approaches than would be dictated by theory. Perhaps as a result, robotic reinforcement learning approaches are often closer to classical optimal control solutions than the methods typically studied in the machine learning literature.

Exploit Domain Structure for Scalability. The grounding of RL in robotics alleviates the general problem of scaling reinforcement learning into high dimensional domains by exploiting the structure of the physical world. Prior knowledge of the behavior of the system and the task is often available. Incorporating even crude models and domain structure into the learning approach (e.g., to approximate gradients) has yielded impressive results(Kolter and Ng, 2009a).

Local Optimality and Controlled State Distributions. Much research in classical reinforcement learning aims at finding value functions that are optimal over the entire state space, which is most likely intractable. In contrast, the success of policy search approaches in robotics relies on their implicit maintenance and controlled change of a state distribution under which the policy can be optimized. Focusing on slowly changing state distributions may also benefit value function methods.

Reward Design. It has been repeatedly demonstrated that reinforcement learning approaches benefit significantly from reward shaping (Ng et al, 1999), and particularly from using rewards that convey a notion of closeness and are not only based on simple binary success or failure. A learning problem is potentially difficult if the reward is sparse, there are significant delays between an action and the associated significant reward, or if the reward is not smooth (i.e., very small changes to the policy lead to a drastically different outcome). In classical reinforcement learning, discrete rewards are often considered, e.g., a small negative reward per time-step and a large positive reward for reaching the goal. In contrast, robotic reinforcement learning approaches often need more physically motivated reward-shaping based on continuous values and consider multi-objective reward functions like minimizing the motor torques while achieving a task.

We hope that these points will help soliciting more new targeted solutions from the reinforcement learning community for robotics.

Movement Templates for Learning of Hitting and Batting

with Katharina Muelling[*], Oliver Kroemer[†], Christoph H. Lampert[‡], and Bernhard Schölkopf[*]

Summary. Hitting and batting tasks, such as tennis forehands, ping-pong strokes, or baseball batting, depend on predictions where the ball can be intercepted and how it can properly be returned to the opponent. These predictions get more accurate over time, hence the behaviors need to be continuously modified. As a result, movement templates with a learned global shape need to be adapted during the execution so that the racket reaches a target position and velocity that will return the ball over to the other side of the net or court. It requires altering learned movements to hit a varying target with the necessary velocity at a specific instant in time. Such a task cannot be incorporated straightforwardly in most movement representations suitable for learning. For example, the standard formulation of the dynamical system based motor primitives (introduced by Ijspeert et al (2002b)) does not satisfy this property despite their flexibility which has allowed learning tasks ranging from locomotion to kendama. In order to fulfill this requirement, we reformulate the Ijspeert framework to incorporate the possibility of specifying a desired hitting point and a desired hitting velocity while maintaining all advantages of the original formulation. We show that the proposed movement template formulation works well in two scenarios, i.e., for hitting a ball on a string with a table tennis racket at a specified velocity and for returning balls launched by a ball gun successfully over the net using forehand movements.

[*] Max Planck Institute for Intelligent Systems, Department of Empirical Inference, Spemannstr. 38, 72076 Tübingen, Germany.
[†] Technische Universität Darmstadt, FB Informatik, FG Intelligent Autonomous Systems, Hochschulstr. 10, 64289 Darmstadt, Germany.
[‡] Institute of Science and Technology Austria, Am Campus 1, IST Austria, A-3400 Klosterneuburg, Austria.

J. Kober and J. Peters, *Learning Motor Skills*,
Springer Tracts in Advanced Robotics 97,
DOI: 10.1007/978-3-319-03194-1_3, © Springer International Publishing Switzerland 2014

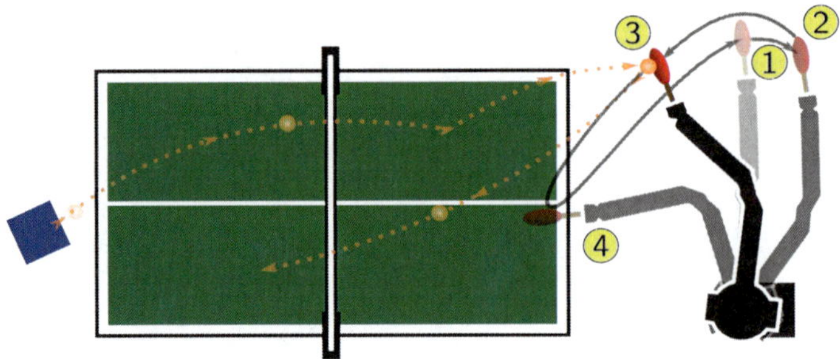

Fig. 3.1 This figure illustrates the different phases of a table tennis stroke. The blue box on the left represents a ping-pong ball launcher, the table is shown in green and the different states of the robot are superposed. A typical racket trajectory is indicated by the dark gray solid line while the orange dashed line represents the ball's trajectory. The robot goes through four stages: it swings back (①→②), it strikes the ball at a virtual hitting point with a goal-oriented velocity at posture ③, it follows the strike through (③→④) and finally returns to the rest posture ①. While the complete arm movement can be modeled with the Ijspeert approach, the reformulation in this paper is required to be able to properly strike the ball.

All experiments were carried out on a Barrett WAM using a four camera vision system.

3.1   Introduction

Learning new skills can frequently be helped significantly by choosing a movement template representation that facilitates the process of acquiring and refining the desired behavior. For example, the work on dynamical systems-based motor primitives (Ijspeert et al, 2002b; Schaal et al, 2003, 2007) has allowed speeding up both imitation and reinforcement learning while, at the same time, making them more reliable. Resulting successes have shown that it is possible to rapidly learn motor primitives for complex behaviors such as tennis swings (Ijspeert et al, 2002b) with only a final target, constrained reaching (Gams and Ude, 2009), drumming (Pongas et al, 2005), biped locomotion (Schaal et al, 2003; Nakanishi et al, 2004) and even in tasks with potential industrial application (Urbanek et al, 2004). Although some of the presented examples, e.g., the tennis swing (Ijspeert et al, 2002b) or the T-ball batting (Peters and Schaal, 2006), are striking movements, these standard motor primitives cannot properly encode a hitting movement. Previous work needed to make simplistic assumptions such as having a static goal (Peters and Schaal, 2006), a learned complex goal function (Peters et al, 2010b) or a stationary goal that could only be lightly touched at the movement's end (Ijspeert et al, 2002b).

Most racket sports require that we hit a non-stationary target at various positions and with various velocities during the execution of a complete striking movement. For example, in table tennis, a typical movement consists of swinging back from a rest postures, hitting the ball at a desired position with a desired orientation and velocity, continuing the swing a bit further and finally returning to the rest posture. See Figure 3.1 for an illustration. Sports sciences literature (Ramanantsoa and Durey, 1994; Muelling and Peters, 2009) indicates that most striking movements are composed of similar phases that only appear to be modified by location and velocity at the interception point for the ball (Schmidt and Wrisberg, 2000; Bootsma and van Wieringen, 1990; Hubbard and Seng, 1954; Tyldesley and Whiting, 1975). These findings indicate that similar motor primitives are being used that are invariant under these external influences similar to the Ijspeert motor primitives (Ijspeert et al, 2002b; Schaal et al, 2003, 2007) being invariant under the modification of the final position, movement amplitude and duration. However, the standard formulation by Ijspeert et al cannot be used properly in this context as there is no possibility to directly incorporate either a via-point or a target velocity (if the duration cannot be adapted as, e.g., for an approaching ball). Hence, a reformulation is needed that can deal with these requirements.

In this paper, we augment the Ijspeert approach (Ijspeert et al, 2002b; Schaal et al, 2003, 2007) of using dynamical systems as motor primitives in such a way that it includes the possibility to set arbitrary velocities at the hitting point without changing the overall shape of the motion or introducing delays that will prevent a proper strike. This modification allows the generalization of learned striking movements, such as hitting and batting, from demonstrated examples. In Section 3.2, we present the reformulation of the motor primitives as well as the intuition behind the adapted approach. We apply the presented method in Section 3.3 where we show two successful examples. First, we test the striking movements primitive in a static scenario of hitting a hanging ball with a table tennis racket and show that the movement generalizes well to new ball locations. After this proof of concept, we take the same movement template representation in order to learn an adaptable forehand for table tennis. The later is tested in the setup indicated by Figure 3.1 where we use a seven degrees of freedom Barrett WAM in order to return balls launched by a ball cannon.

3.2   Movement Templates for Learning to Strike

Ijspeert et al (2002b) suggested to use a dynamical system approach in order to represent both discrete point-to-point movements as well as rhythmic motion with motor primitives. This framework ensures the stability of the movement[1], allows the representation of arbitrarily shaped movements through

---

[1] Note that the dynamical systems motor primitives ensure the stability of the movement generation but cannot guarantee the stability of the movement execution (Ijspeert et al, 2002b; Schaal et al, 2007).

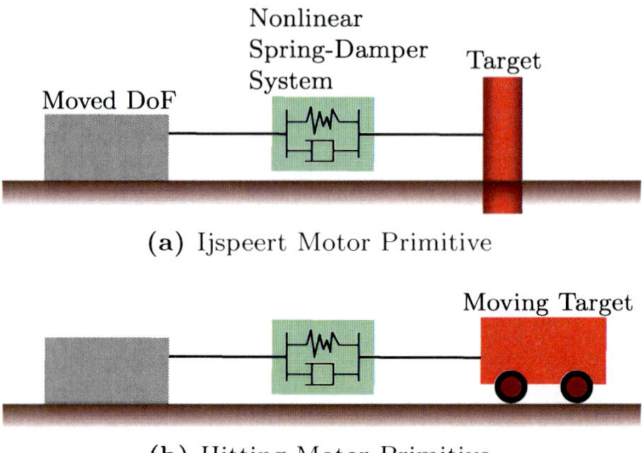

(a) Ijspeert Motor Primitive

(b) Hitting Motor Primitive

Fig. 3.2 In this figure, we convey the intuition of the presented reactive templates for learning striking movements. The Ijspeert formulation can be seen as a nonlinear spring damper system that pulls a degree of freedom to a stationary goal while exhibiting a learned movement shape. The presented approach allows hitting a target with a specified velocity without replanning if the target is adapted and, in contrast to the Ijspeert formulation, can be seen as a degree of freedom pulled towards a moving goal.

the primitive's policy parameters, and these parameters can be estimated straightforwardly by locally weighted regression. In the discrete case, these primitives can be modified through their meta-parameters in order to adapt to the final goal position, the movement amplitude or the duration of the movement. The resulting movement can start from arbitrary positions and velocities and go to arbitrary final positions while maintaining the overall shape of the trajectory. In Section 3.2.1, we review the most current version of this approach based on (Schaal et al, 2007).

However, as outlined in Section 3.1, this formulation of the motor primitives cannot be used straightforwardly in racket sports as incorporating a desired virtual hitting point (Ramanantsoa and Durey, 1994; Muelling and Peters, 2009) (consisting of a desired target position and velocity) cannot be achieved straightforwardly. For example, in table tennis, fast forehand movements need to hit a ball at a pre-specified speed, hitting time and a continuously adapted location. In the original Ijspeert formulation, the goal needs to be determined at the start of the movement and at approximately zero velocity as in the experiments in (Ijspeert et al, 2002b); a via-point target can only be hit properly by modifying either the policy parameters (Peters and Schaal, 2006) or, indirectly, by modifying the goal parameters (Peters et al, 2010b). Hence, such changes of the target can only be achieved by drastically changing the shape of the trajectory and duration.

As an alternative, we propose a modified version of Ijspeert's original framework that overcomes this limitation and is particularly well-suited for

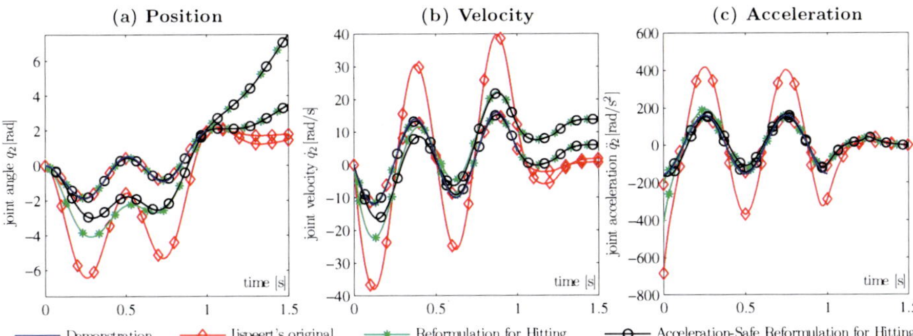

Fig. 3.3 Target velocity adaptation is essential for striking movements. This figure illustrates how different versions of the dynamical system based motor primitives are affected by a change of the target velocity. Here, an artificial training example (i.e., $q = 2t^2 + \cos(4t\pi) - 1$) is generated. After learning, all motor primitive formulations manage to reproduce the movements accurately from the training example for the same target velocity and cannot be distinguished. When the target velocity is tripled, this picture changes drastically. For Ijspeert's original model the amplitude modifier **a** had to be increased to yield the desired velocity. The increased amplitude of the trajectory is clearly visible for the positions and even more drastic for the velocities and accelerations. The reformulations presented in this paper, stay closer to the movement shape and amplitude. Particularly the velocities and accelerations exhibit that the new approach allows much better generalizing of the learned behavior. This figure furthermore demonstrates how a large initial step in acceleration appears for Ijspeert's original model (and the reformulation for hitting) even if a transformation function is used to partially suppress it for the training example.

striking movements. This modification allows setting both a hitting point and a striking velocity while maintaining the desired duration and the learned shape of the movement. Online adaptation of these meta-parameters is possible and, hence, it is well-suited for learning racket sports such as table tennis as discussed in Chapter 5. The basic intuition behind the modified version is similar to the one of Ijspeert's primitives, i.e., both assume that the controlled degree of freedom is connected to a specific spring damper system; however, the approach presented here allows overcoming previous limitations by assuming a connected moving target, see Figure 3.2. The resulting approach is explained in Section 3.2.2.

A further drawback of the Ijspeert motor primitives (Ijspeert et al, 2002b; Schaal et al, 2003, 2007) is that, when generalizing to new targets, they tend to produce large accelerations early in the movement. Such an acceleration peak may not be well-suited for fast movements and can lead to execution problems due to physical limitations; Figure 3.4 illustrates this drawback. In Section 3.2.3, we propose a modification that alleviates this shortcoming.

3.2.1 Discrete Movement Primitives

While the original formulation in (Ijspeert et al, 2002b; Schaal et al, 2003) for discrete dynamical systems motor primitives used a second-order system to represent the phase z of the movement, this formulation has proven to be unnecessarily complicated in practice. Since then, it has been simplified and, in (Schaal et al, 2007), it was shown that a single first order system suffices

$$\dot{z} = -\tau\alpha_z z.$$

This canonical system has the time constant $\tau = 1/T$ where T is the duration of the motor primitive, a parameter α_z which is chosen such that $z \approx 0$ at T to ensure that the influence of the transformation function, shown in Equation (3.2.1), vanishes. Subsequently, the internal state $\mathbf{x}$ of a second system is chosen such that positions $\mathbf{q}$ of all degrees of freedom are given by $\mathbf{q} = \mathbf{x}_1$, the velocities $\dot{\mathbf{q}}$ by $\dot{\mathbf{q}} = \tau\mathbf{x}_2 = \dot{\mathbf{x}}_1$ and the accelerations $\ddot{\mathbf{q}}$ by $\ddot{\mathbf{q}} = \tau\dot{\mathbf{x}}_2$. Under these assumptions, the learned dynamics of Ijspeert motor primitives can be expressed in the following form

$$\dot{\mathbf{x}}_2 = \tau\alpha_x\left(\beta_x\left(\mathbf{g} - \mathbf{x}_1\right) - \mathbf{x}_2\right) + \tau\mathbf{A}\mathbf{f}\left(z\right), \tag{3.1}$$

$$\dot{\mathbf{x}}_1 = \tau\mathbf{x}_2.$$

This set of differential equations has the same time constant τ as the canonical system, parameters α_x, β_x set such that the system is critically damped, a goal parameter $\mathbf{g}$, a transformation function $\mathbf{f}$ and an amplitude matrix $\mathbf{A} = \mathrm{diag}(a_1, a_2, \ldots, a_n)$, with the amplitude modifier $\mathbf{a} = [a_1, a_2, \ldots, a_n]$. In (Schaal et al, 2007), they use $\mathbf{a} = \mathbf{g} - \mathbf{x}_1^0$ with the initial position $\mathbf{x}_1^0$, which ensures linear scaling. Alternative choices are possibly better suited for specific tasks, see e.g., (Park et al, 2008). The transformation function $\mathbf{f}\left(z\right)$ alters the output of the first system, in Equation (3.2.1), so that the second system, in Equation (3.1), can represent complex nonlinear patterns and it is given by

$$\mathbf{f}\left(z\right) = \sum_{i=1}^{N}\psi_i\left(z\right)\mathbf{w}_i z.$$

Here, $\mathbf{w}_i$ contains the ith adjustable parameter of all degrees of freedom, N is the number of parameters per degree of freedom, and $\psi_i(z)$ are the corresponding weighting functions (Schaal et al, 2007). Normalized Gaussian kernels are used as weighting functions given by

$$\psi_i\left(z\right) = \frac{\exp\left(-h_i\left(z - c_i\right)^2\right)}{\sum_{j=1}^{N}\exp\left(-h_j\left(z - c_j\right)^2\right)}.$$

These weighting functions localize the interaction in phase space using the centers c_i and widths h_i. Note that the degrees of freedom (DoF) are usually all modeled as independent in Equation (3.1). All DoFs are synchronous as the dynamical systems for all DoFs start at the same time, have the same duration, and the shape of the movement is generated using the transformation $\mathbf{f}(z)$ in Equation (3.2.1). This transformation function is learned as a function of the shared canonical system in Equation (3.2.1).

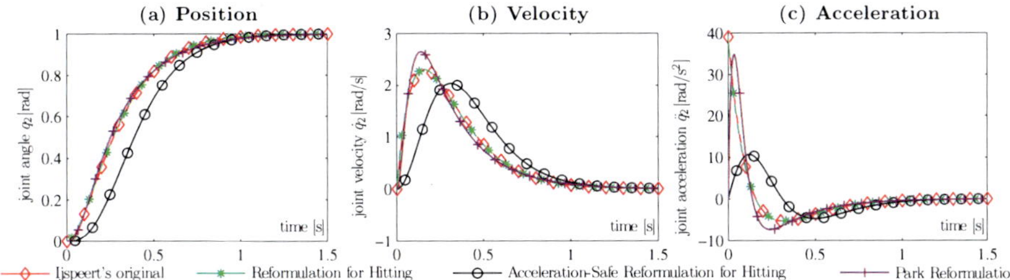

Fig. 3.4 An important aspect of the Ijspeert framework is that such primitives are guaranteed to be stable and, hence, safe for learning. A problem of the regular formulation highly unevenly distributed acceleration with a jump at the beginning of the movement of its unaltered dynamics. These high accelerations affect the movement when the behavior is either generalized to new goals or when during trial-and-error learning where the initial parameters are small. Some of these problem have previously been noted by Park et al (2008), and are particularly bad in the context of fast striking movements. Here, we compare the different formulations with respect to their acceleration in the unaltered dynamics case (i.e., $\mathbf{w} = \mathbf{0}$). For a better comparison, we set the goal velocity to zero ($\dot{\mathbf{g}} = 0$). The Ijspeert formulation clearly shows the problem with the large acceleration, as does the reformulation for hitting (with a hitting speed of $\dot{\mathbf{g}} = 0$ both are identical). While the Park modification starts without the jump in acceleration, it requires almost as high accelerations shortly afterwards. The acceleration-safe reformulation for hitting also starts out without a step in acceleration and does not require huge accelerations.

One of the biggest advantages of this motor primitive framework (Ijspeert et al, 2002a; Schaal et al, 2007) is that the second system in Equation (3.1), is linear in the shape parameters $\boldsymbol{\theta}$. Therefore, these parameters can be obtained efficiently, and the resulting framework is well-suited for imitation (Section 3.2.4) and reinforcement learning (Chapter 4). Additional feedback terms can be added as shown in (Schaal et al, 2007; Kober et al, 2008; Park et al, 2008).

3.2.2   *Adapting the Motor Primitives for Striking Movements*

The regular formulation (Ijspeert et al, 2002b; Schaal et al, 2003, 2007) which was reviewed in Section 3.2.1, allows to change the initial position $\mathbf{x}_1^0$ and goal position $\mathbf{g}$ (which corresponds to the target, i.e., the position at the end of the

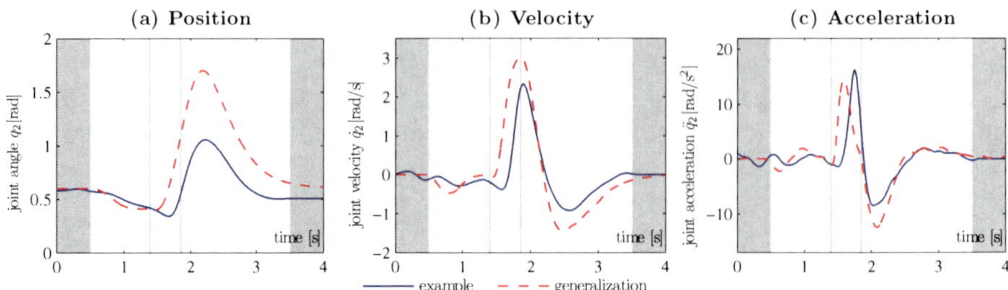

Fig. 3.5 This figure demonstrates the generalization of an imitated behavior to a different target that is $15cm$ away from the original target. Note that this trajectory is for a static target, hence the slow motion. The depicted degree of freedom (DoF) is shoulder adduction-abduction (i.e., the second DoF). The solid gray bars indicate the time before and after the main movement, the gray dashed lines indicate the phase borders also depicted in Figure 3.1 and the target is hit at the second border.

movement at time T) of the motor primitive while maintaining the overall shape of the movement determined by the parameters $\mathbf{w}_i$. For disturbed initial conditions, the attractor dynamics that pull the motor primitive to the trained behavior and it is guaranteed to finally reach to the goal position $\mathbf{g}$, see (Ijspeert et al, 2002b). However, the formulation above only considers the case of a final goal with a favored velocity of $\dot{\mathbf{x}}_1(T) = 0$ at the goal $\mathbf{g}$ and final time T. However, using the transformation function $\mathbf{f}(z)$ in Equation (3.2.1), it can be forced to arbitrary final velocities by changing the shape parameters of the movement. As the last basis function in the transformation function $\mathbf{f}(z)$ decays almost to zero at time T the active parameters, the last parameter $\mathbf{w}_N$ needs to be over-proportionally large. If the motor primitive is trained with $\dot{\mathbf{x}}_1(T) = 0$ it simply rests at $\mathbf{x}_1 = \mathbf{g}$ if it runs for longer than T. However, large $\mathbf{w}_N$ often cause overshooting in $\mathbf{x}_1$ and the trajectory is subsequently pulled back to the final position $\mathbf{g}$ only using the linear attractor dynamics in Equation (3.1) which may not be suitable for a given task. The goal velocity $\dot{\mathbf{x}}_1(T)$ can only be changed either by scaling the duration of the movement T or with the amplitude modifier $\mathbf{a}$; however a mapping of $\mathbf{g}$ and $\dot{\mathbf{x}}_1(T)$ to $\mathbf{a}$ has to be established first. The main downsides of these approaches, respectively, are that either the total duration is changed (which makes the interception of a table tennis ball hard) or that $\mathbf{a}$ modifies the whole motion including shape and amplitude (which causes undesired movements and often requires overly strenuous movements in table tennis). These effects are illustrated in Figure 3.3. Note, if the goal is constantly adapted as in table tennis (where the ball trajectory is not certain until the ball has bounced on the table the last time), these effects will produce significantly stronger undesired effects and, possibly, unstable behavior.

As an alternative for striking movements, we propose a modification of the dynamical system based motor primitives that allows us to directly specify the desired $\dot{\mathbf{x}}_1(T)$ while maintaining the duration of the movement and having the possibility to change the amplitude of the motion independently. For doing so, we introduce a moving goal and include the desired final velocity in Equation (3.1). We use a linearly moving goal but other choices may be better suited for different tasks. This reformulation results in the following equations for the learned dynamics

$$\dot{\mathbf{x}}_2 = \tau\alpha_g \left(\beta_g \left(\mathbf{g}_m - \mathbf{x}_1\right) + \frac{(\dot{\mathbf{g}} - \dot{\mathbf{x}}_1)}{\tau} \right) + \tau\mathbf{A}\mathbf{f}, \qquad (3.2)$$

$$\dot{\mathbf{x}}_1 = \tau\mathbf{x}_2,$$

$$\mathbf{g}_m = \mathbf{g}_m^0 - \dot{\mathbf{g}}\frac{\ln{(z)}}{\tau\alpha_h}, \qquad (3.3)$$

where $\dot{\mathbf{g}}$ is the desired final velocity, $\mathbf{g}_m$ is the moving goal and the initial position of the moving goal $\mathbf{g}_m^0 = \mathbf{g} - \tau\dot{\mathbf{g}}$ ensures that $\mathbf{g}_m\left(T\right) = \mathbf{g}$. The term $-\ln{(z)}/(\tau\alpha_h)$ is proportional to the time if the canonical system in Equation (3.2.1) runs unaltered; however, adaptation of z allows the straightforward adaptation of the hitting time. If $\dot{\mathbf{g}} = \mathbf{0}$, this formulation is exactly the same as the original formulation. The imitation learning approach mentioned in Section 3.2.1 can be adapted straightforwardly to this formulation. Figure 3.3 illustrates how the different approaches behave when forced to achieve a specified desired final velocity.

3.2.3   Safer Dynamics for Generalization

Generalizing fast movements such as a forehand in table tennis can become highly dangerous if the primitive requires exceedingly high accelerations or has large jumps in the acceleration (e.g., the fastest table tennis moves that we have executed on our WAM had a peak velocity of $7m/s$ and $10g$ maximum acceleration). Hence, the initial jump in acceleration often observed during the execution of the Ijspeert primitives may lead to desired accelerations that a physical robot cannot properly execute, and may even cause damage to the robot system. In the following, we will discuss several sources of these acceleration jumps and how to overcome them. If the dynamics are not altered by the transformation function, i.e., $\mathbf{w} = \mathbf{0}$, the highest acceleration during the original Ijspeert motor primitive occurs at the very first time-step and then decays rapidly. If the motor primitives are properly initialized by imitation learning, the transformation function will cancel this initial acceleration, and, thus, this usually does not pose a problem in the absence of generalization. However, when changing the amplitude $\mathbf{a}$ of the motion (e.g., in order to achieve a specific goal velocity) the transformation function will over- or undercompensate for this initial acceleration jump. The adaptation proposed in Section 3.2.2 does not require a change in amplitude, but suffers from a related shortcoming, i.e., changing the goal velocity also changes the initial position of the goal, thus results in a similar jump in acceleration that needs to be compensated. Using the motor primitives with an initial velocity that differs from the one used during imitation learning has the same effect. Figures 3.3 and 3.4 illustrate these initial steps in acceleration for various motor primitive formulations. As an alternative, we propose to gradually activate the attractor dynamics of the motor primitive (e.g., by reweighting them using the output of the canonical system). When combined, these two adaptations result in

$$\dot{\mathbf{x}}_2 = (1 - z)\,\tau\alpha_g\left(\beta_g\left(\mathbf{g}_m - \mathbf{x}_1\right) + \frac{(\dot{\mathbf{g}} - \dot{\mathbf{x}}_1)}{\tau}\right) + \tau\mathbf{A}\mathbf{f}.$$

Surprisingly, after this modification the unaltered dynamics (i.e., where $\mathbf{w} = \mathbf{0}$ and, hence, $\tau\mathbf{A}\mathbf{f}(z) = \mathbf{0}$) result in trajectories that roughly resemble a minimum jerk movements and, hence, look very similar to human movements. Exactly as for the Ijspeert formulation, we can arbitrarily shape the behavior by learning the weights of the transformation function. Note that (Park et al, 2008) also introduced a similar modification canceling the initial acceleration caused by the offset between initial and goal position; however, their approach cannot deal with a deviating initial velocity.

The proposed acceleration jump compensation also yields smoother movements during the adaptation of the hitting point as well as smoother transitions if motor primitives are sequenced. The later becomes particularly important when the preceding motor primitive has a significantly different velocity than during training (by imitation learning) or if it is terminated early due to external events. All presented modifications are compatible with the imitation learning approach discussed in Section 3.2.1 and the adaptation is straightforward. Figures 3.3 and 3.4 show how the presented modifications overcome the problems with the initial jumps in acceleration.

3.2.4 *Imitation Learning*

In the following chapters we use imitation learning from a single example to generate a sensible initial policy. This step can be performed efficiently in the context of dynamical systems motor primitives in both the original and adapted forms, as the transformation function Equation (3.2.1) is linear in its parameters. As a result, we can choose the weighted squared error (WSE)

$$\mathrm{WSE}_n = \sum_{t=1}^{T}\psi_t^n\left(f_t^{\mathrm{ref}} - z_t\theta^n\right)^2$$

as cost function and minimize it for all parameter vectors θ^n with $n \in \{1, 2, \ldots, N\}$. Here, the corresponding weighting functions are denoted by ψ_t^n and the basis function by z_t. The reference or target signal f_t^{ref} is the desired transformation function and t indicates the time-step of the sample. The error in Equation (3.2.4) can be rewritten in matrix form as

$$\mathrm{WSE}_n = \left(\mathbf{f}^{\mathrm{ref}} - \mathbf{Z}\theta^n\right)^{\mathrm{T}}\mathbf{\Psi}\left(\mathbf{f}^{\mathrm{ref}} - \mathbf{Z}\theta^n\right)$$

(a) Demonstration by a Human Instructor

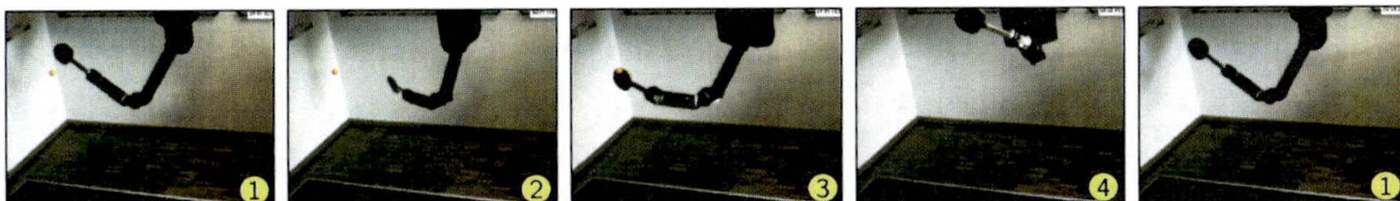

(b) Example: Reproduction for Hitting a Stationary Ball

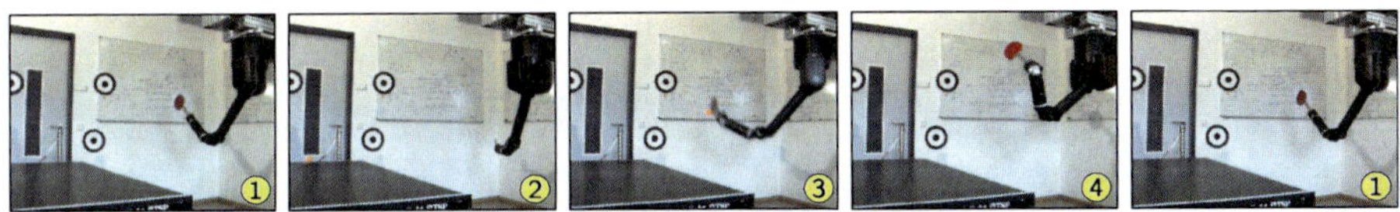

(c) Application: Returning Balls launched by a Ball Gun

Fig. 3.6 This figure presents a hitting sequence from the demonstration, a generalization on the robot with a ball attached by a string as well as a generalization hitting a ball shot by a ping-pong ball launcher. The demonstration and the flying ball generalization are captured by a $25Hz$ video camera, the generalization with the attached ball is captured with $200Hz$ through our vision system. From left to right the stills represent: rest posture, swing-back posture, hitting point, swing-through and rest posture. The postures (①-④) are the same as in Figure 3.2.

with $\mathbf{f}^{\mathrm{ref}}$ containing the values of f_t^{ref} for all time-steps t, $\mathbf{\Psi} = \mathrm{diag}\,(\psi_1^n, \ldots, \psi_t^n, \ldots, \psi_T^n)$, and $[\mathbf{Z}]_t = z_t$. As a result, we have a standard locally-weighted linear regression problem that is straightforward to solve and yields the unbiased parameter estimator

$$\theta^n = \left(\mathbf{Z}^{\mathrm{T}}\mathbf{\Psi}\mathbf{Z}\right)^{-1}\mathbf{Z}^{\mathrm{T}}\mathbf{\Psi}\mathbf{f}^{\mathrm{ref}}.$$

This approach was originally suggested for imitation learning by Ijspeert et al (2002b). Estimating the parameters of the dynamical system is slightly more difficult; the duration of the movement is extracted using motion detection and the time-constant is set accordingly.

3.3   Robot Evaluation

In this section, we evaluate the presented reactive templates for representing, learning and executing forehands in the setting of table tennis. For doing so, we evaluate our representation for striking movements first on hitting a hanging ball in Section 3.3.1 and, subsequently, in the task of returning a ball served by a ball launcher presented in Section 3.3.2.

When hitting a ping-pong ball that is hanging from the ceiling, the task consists of hitting the ball with an appropriate desired Cartesian velocity and orientation of the paddle. Hitting a ping-pong ball shot by a ball launcher requires predicting the ball's future positions and velocities in order to choose an interception point. The latter is only sufficiently accurate after the ball has hit the table for the last time. This short reaction time underlines that the movement templates can be adapted during the trajectory under strict time limitations when there is no recovery from a bad generalization, long replanning or inaccurate movements.

3.3.1 Generalizing Forehands on Static Targets

As a first experiment, we evaluated how well this new formulation of hitting primitives generalizes forehand movements learned from imitation as shown in Figure 3.6 (a). First, we collected arm, racket and ball trajectories for imitation learning using the 7 DoF Barrett WAM robot as an haptic input device for kinesthetic teach-in where all inertial forces and gravity were compensated. In the second step, we employ this data to automatically extract the duration of the striking movement, the duration of the individual phases as well as the Cartesian target velocity and orientation of the racket when hitting the ball. We employ a model (as shown in Section 3.2) that has phases for swinging back, hitting and going to a rest posture. Both the phase for swing-back and return-to-home phases will go into intermediary still phases while the hitting phase goes through a target point with a pre-specified target velocity. All phases can only be safely executed due to the "safer dynamics" which we introduced in Section 3.2.3.

In this experiment, the ball is a stationary target and detected by a stereo camera setup. Subsequently, the supervisory level proposed in (Muelling and Peters, 2009) determines the hitting point and the striking velocity in configuration space. The motor primitives are adjusted accordingly and executed on the robot in joint-space using an inverse dynamics control law. The robot successfully hits the ball at different positions within a diameter of approximately $1.2m$ if kinematically feasible. The adaptation for striking movements achieves the desired velocities and the safer dynamics allow generalization to a much larger area while successfully removing the possibly large accelerations at the transitions between motor primitives. See Figure 3.5 for a comparison of the training example and the generalized motion for one degree of freedom and Figure 3.6 (b) for a few frames from a hit of a static ball.

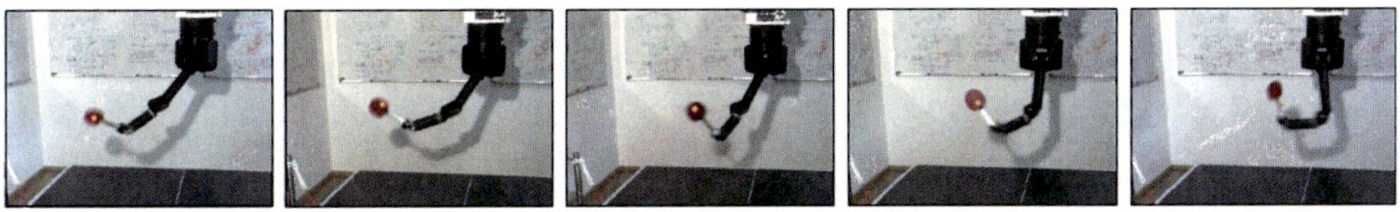

Fig. 3.7 Generalization to various targets (five different forehands at posture ③) are shown approximately when hitting the ball

3.3.2 Playing against a Ball Launcher

This evaluation adds an additional layer of complexity as the hitting point and the hitting time has to be estimated from the trajectory of the ball and continuously adapted as the hitting point cannot be reliably determined until the ball has bounced off the table for the last time. In this setting, the ball is tracked by two overlapping high speed stereo vision setups with $200Hz$ cameras. In order to obtain better estimates of the current position and to calculate the velocities, the raw 3D positions are filtered by a specialized Kalman filter (Kalman, 1960) that takes contacts of the ball with the table and the racket into account (Muelling and Peters, 2009). When used as a Kalman predictor, we can again determine the target point for the primitive with a pre-specified target velocity with the method described in (Muelling and Peters, 2009). The results obtained for the still ball generalize well from the static ball to the one launched by a ball launcher at $3m/s$ which are returned at speeds up to $8m/s$. A sequence of frames from the attached video is shown in Figure 3.6. The plane of possible virtual hitting points again has a diameter of roughly $1m$ as shown in Figure 3.7. The modified motor primitives generated movements with the desired hitting position and velocity. The robot hit the ball in the air in approx. 95% of the trials. However, due to a simplistic ball model and execution inaccuracies the ball was often not properly returned on the table. Please see the videos accompanying this chapter `http://www.robot-learning.de/Research/HittingMPs`.

Note that our results differ significantly from previous approaches as we use a framework that allows us to learn striking movements from human demonstrations unlike previous work in batting (Senoo et al, 2006) and table tennis (Andersson, 1988). Unlike baseball which only requires four degrees of freedom (as, e.g., in (Senoo et al, 2006) who used a 4 DoF WAM arm in a manually coded high speed setting), and previous work in table tennis (which had only low-inertia, was overpowered and had mostly prismatic joints (Andersson, 1988; Fässler et al, 1990; Matsushima et al, 2005)), we use a full seven degrees of freedom revolutionary joint robot and, thus, have to deal with larger inertia as the wrist adds roughly $2.5kg$ weight at the elbow. Hence, it was essential to train trajectories by imitation learning that distribute the torques well over the redundant joints as the human teacher was suffering from the same constraints.

3.4   Conclusion

In this paper, we rethink previous work on dynamic systems motor primitive (Ijspeert et al, 2002b; Schaal et al, 2003, 2007) in order to obtain movement templates that can be used reactively in batting and hitting sports. This reformulation allows to change the target velocity of the movement while maintaining the overall duration and shape. Furthermore, we present a modification that overcomes the problem of an initial acceleration step which is particularly important for safe generalization of learned movements. Our adaptations retain the advantages of the original formulation and perform well in practice. We evaluate this novel motor primitive formulation first in hitting a stationary table tennis ball and, subsequently, in returning ball served by a ping pong ball launcher. In both cases, the novel motor primitives manage to generalize well while maintaining the features of the demonstration. This new formulation of the motor primitives can hopefully be used together with *meta-parameter learning* (Chapter 5) in a *mixture of motor primitives* (Muelling et al, 2010) in order to create a complete framework for learning tasks like table tennis autonomously.

4

Policy Search for Motor Primitives in Robotics

Summary. Many motor skills in humanoid robotics can be learned using parametrized motor primitives. While successful applications to date have been achieved with imitation learning, most of the interesting motor learning problems are high-dimensional reinforcement learning problems. These problems are often beyond the reach of current reinforcement learning methods. In this chapter, we study parametrized policy search methods and apply these to benchmark problems of motor primitive learning in robotics. We show that many well-known parametrized policy search methods can be derived from a general, common framework. This framework yields both policy gradient methods and expectation-maximization (EM) inspired algorithms. We introduce a novel EM-inspired algorithm for policy learning that is particularly well-suited for dynamical system motor primitives. We compare this algorithm, both in simulation and on a real robot, to several well-known parametrized policy search methods such as episodic REINFORCE, 'Vanilla' Policy Gradients with optimal baselines, episodic Natural Actor Critic, and episodic Reward-Weighted Regression. We show that the proposed method out-performs them on an empirical benchmark of learning dynamical system motor primitives both in simulation and on a real robot. We apply it in the context of motor learning and show that it can learn a complex Ball-in-a-Cup task on a real Barrett WAM robot arm.

4.1 Introduction

To date, most robots are still taught by a skilled human operator either via direct programming or a teach-in. Learning approaches for automatic task acquisition and refinement would be a key step for making robots progress towards autonomous behavior. Although imitation learning can make this task more straightforward, it will always be limited by the observed demonstrations. For many motor learning tasks, skill transfer by imitation learning

J. Kober and J. Peters, *Learning Motor Skills*,
Springer Tracts in Advanced Robotics 97,
DOI: 10.1007/978-3-319-03194-1_4, © Springer International Publishing Switzerland 2014

is prohibitively hard given that the human teacher is not capable of conveying sufficient task knowledge in the demonstration. In such cases, reinforcement learning is often an alternative to a teacher's presentation, or a means of improving upon it. In the high-dimensional domain of anthropomorphic robotics with its continuous states and actions, reinforcement learning suffers particularly from the curse of dimensionality. However, by using a task-appropriate policy representation and encoding prior knowledge into the system by imitation learning, local reinforcement learning approaches are capable of dealing with the problems of this domain. Policy search (also known as policy learning) is particularly well-suited in this context, as it allows the usage of domain-appropriate pre-structured policies (Toussaint and Goerick, 2007), the straightforward integration of a teacher's presentation (Guenter et al, 2007; Peters and Schaal, 2006) as well as fast online learning (Bagnell et al, 2003; Ng and Jordan, 2000; Hoffman et al, 2007). Recently, policy search has become an accepted alternative of value-function-based reinforcement learning (Bagnell et al, 2003; Strens and Moore, 2001; Kwee et al, 2001; Peshkin, 2001; El-Fakdi et al, 2006; Taylor et al, 2007) due to many of these advantages.

In this chapter, we will introduce a policy search framework for episodic reinforcement learning and show how it relates to policy gradient methods (Williams, 1992; Sutton et al, 1999; Lawrence et al, 2003; Tedrake et al, 2004; Peters and Schaal, 2006) as well as expectation-maximization (EM) inspired algorithms (Dayan and Hinton, 1997; Peters and Schaal, 2007). This framework allows us to re-derive or to generalize well-known approaches such as episodic REINFORCE (Williams, 1992), the policy gradient theorem (Sutton et al, 1999; Peters and Schaal, 2006), the episodic Natural Actor Critic (Peters et al, 2003, 2005), and an episodic generalization of the Reward-Weighted Regression (Peters and Schaal, 2007). We derive a new algorithm called Policy Learning by Weighting Exploration with the Returns (PoWER), which is particularly well-suited for the learning of trial-based tasks in motor control.

We evaluate the algorithms derived from this framework to determine how they can be used for refining parametrized policies in robot skill learning. To address this problem, we follow a methodology suitable for robotics where the policy is first initialized by imitation learning and, subsequently, the policy search algorithm is used for self-improvement. As a result, we need a suitable representation in order to apply this approach in anthropomorphic robot systems. In imitation learning, a particular kind of motor control policy has been very successful, which is known as dynamical system motor primitives (Ijspeert et al, 2002b,a; Schaal et al, 2003, 2007). In this approach, dynamical systems are used to encode a control policy suitable for motor tasks. The representation is linear in the parameters; hence, it can be learned straightforwardly from demonstrations. Such dynamical system motor primitives can represent both point-to-point and rhythmic behaviors. We focus on the point-to-point variant which is suitable for representing single-stroke, episodic behaviors. As a result, they are particularly well-suited for episodic policy search.

We show that all presented algorithms work sufficiently well when employed in the context of learning dynamical system motor primitives in different benchmark and application settings. We compare these methods on the two benchmark problems from (Peters and Schaal, 2006) for dynamical system motor primitives learning, the Underactuated Swing-Up (Atkeson, 1994) robotic benchmark problem, and the Casting task. Using entirely different parametrizations, we evaluate policy search methods on the mountain-car benchmark (Sutton and Barto, 1998) and the Tetherball Target Hitting task. On the mountain-car benchmark, we additionally compare to a value function based approach. The method with the best performance, PoWER, is evaluated on the complex task of Ball-in-a-Cup (Sumners, 1997). Both the Underactuated Swing-Up as well as Ball-in-a-Cup are achieved on a real Barrett WAM robot arm. Please also refer to the videos at `http://www.robot-learning.de/Research/ReinforcementLearning`. For all real robot experiments, the presented movement is learned by imitation from a kinesthetic demonstration, and the Barrett WAM robot arm subsequently improves its behavior by reinforcement learning.

4.2   Policy Search for Parametrized Motor Primitives

Our goal is to find reinforcement learning techniques that can be applied in robotics in the context of learning high-dimensional motor control tasks. We first introduce the required notation for the derivation of the reinforcement learning framework in Section 4.2.1. We discuss the problem in the general setting of reinforcement learning using a generalization of the approach in (Dayan and Hinton, 1997; Attias, 2003; Peters and Schaal, 2007). We extend the existing approach to episodic reinforcement learning for continuous states, in a manner suitable for robotics.

We derive a new expectation-maximization (EM) inspired algorithm (Dempster et al, 1977) called Policy Learning by Weighting Exploration with the Returns (PoWER) in Section 4.2.3 and show how the general framework is related to policy gradient methods and the Reward-Weighted Regression method in Section 4.2.2.

4.2.1   Problem Statement and Notation

In this chapter, we treat motor primitive learning problems in the framework of reinforcement learning (Sutton and Barto, 1998) with a strong focus on the episodic case. At time t, there is an actor in a state $\mathbf{s}_t$ that chooses an action $\mathbf{a}_t$ according to a stochastic policy $\pi(\mathbf{a}_t|\mathbf{s}_t, t)$. Such a policy is a probability distribution over actions given the current state and time. The stochastic formulation allows a natural incorporation of exploration, and the optimal time-invariant policy has been shown to be stochastic in the case of hidden state variables (Sutton et al, 1999; Jaakkola et al, 1993). Upon the completion of the action, the actor transfers to a state $\mathbf{s}_{t+1}$ and receives a reward r_t. As we are interested in learning complex motor tasks consisting of a single stroke (Schaal et al, 2007), we focus on finite horizons of length T with episodic restarts and learn the optimal parametrized, stochastic policy for such episodic reinforcement learning problems (Sutton and Barto, 1998). We assume an explorative parametrized policy π with parameters $\boldsymbol{\theta} \in \mathbb{R}^n$. In Section 4.3.1, we discuss how the dynamical system motor primitives (Ijspeert et al, 2002b,a; Schaal et al, 2003, 2007) can be employed in this setting. In this section, we will keep most derivations sufficiently general such that they are transferable to various other parametrized policies that are linear in the parameters.

The general goal in reinforcement learning is to optimize the *expected return* of the policy π with parameters $\boldsymbol{\theta}$ defined by

$$ J(\boldsymbol{\theta}) = \int_{\mathbb{T}} p_{\boldsymbol{\theta}}(\boldsymbol{\tau}) R(\boldsymbol{\tau}) d\boldsymbol{\tau}, $$

where $\mathbb{T}$ is the set of all possible paths. A rollout $\boldsymbol{\tau} = [\mathbf{s}_{1:T+1}, \mathbf{a}_{1:T}]$, also called path, episode or trial, denotes a series of states $\mathbf{s}_{1:T+1} = [\mathbf{s}_1, \mathbf{s}_2, \ldots, \mathbf{s}_{T+1}]$ and actions $\mathbf{a}_{1:T} = [\mathbf{a}_1, \mathbf{a}_2, \ldots, \mathbf{a}_T]$. The probability of rollout $\boldsymbol{\tau}$ is denoted by $p_{\boldsymbol{\theta}}(\boldsymbol{\tau})$, while $R(\boldsymbol{\tau})$ refers to its aggregated return. Using the standard Markov assumption and additive accumulated rewards, we can write

$$ p_{\boldsymbol{\theta}}(\boldsymbol{\tau}) = p(\mathbf{s}_1)\prod_{t=1}^{T} p(\mathbf{s}_{t+1}|\mathbf{s}_t, \mathbf{a}_t)\pi(\mathbf{a}_t|\mathbf{s}_t, t), \tag{4.1} $$
$$ R(\boldsymbol{\tau}) = T^{-1}\sum_{t=1}^{T} r(\mathbf{s}_t, \mathbf{a}_t, \mathbf{s}_{t+1}, t), $$

where $p(\mathbf{s}_1)$ denotes the initial state distribution, $p(\mathbf{s}_{t+1}|\mathbf{s}_t, \mathbf{a}_t)$ the next state distribution conditioned on the last state and action, and $r(\mathbf{s}_t, \mathbf{a}_t, \mathbf{s}_{t+1}, t)$ denotes the immediate reward.

While episodic Reinforcement Learning (RL) problems with finite horizons are common in both human Wulf (2007) and robot motor control problems, few methods exist in the RL literature. Examples are episodic REIN-FORCE (Williams, 1992), the episodic Natural Actor Critic eNAC (Peters et al, 2003, 2005) and model-based methods using differential dynamic programming (Atkeson, 1994).

4.2.2 Episodic Policy Learning

In this section, we discuss episodic reinforcement learning in policy space, which we will refer to as Episodic Policy Learning. We first discuss the lower bound on the expected return as suggested in (Dayan and Hinton, 1997) for guaranteeing that policy update steps are improvements. In (Dayan and Hinton, 1997; Peters and Schaal, 2007) only the immediate reward case is discussed; we extend this framework to episodic reinforcement learning. Subsequently, we derive a general update rule, which yields the policy gradient theorem (Sutton et al, 1999), a generalization of the reward-weighted regression (Peters and Schaal, 2007), as well as the novel Policy learning by Weighting Exploration with the Returns (PoWER) algorithm.

Bounds on Policy Improvements

Unlike in reinforcement learning, other branches of machine learning have focused on maximizing lower bounds on the cost functions, which often results in expectation-maximization (EM) algorithms (McLachan and Krishnan, 1997). The reasons for this preference also apply in policy learning: if the lower bound also becomes an equality for the sampling policy, we can guarantee that the policy will be improved by maximizing the lower bound. Results from supervised learning can be transferred with ease. First, we generalize the scenario suggested by Dayan and Hinton (1997) to the episodic case. Here, we generate rollouts τ using the current policy with parameters $\boldsymbol{\theta}$, which we then weight with the returns $R(\tau)$, and subsequently match it with a new policy parametrized by $\boldsymbol{\theta}'$. This matching of the success-weighted path distribution is equivalent to minimizing the Kullback-Leibler divergence $D(p_{\boldsymbol{\theta}}(\tau)R(\tau)\|p_{\boldsymbol{\theta}'}(\tau))$ between the new path distribution $p_{\boldsymbol{\theta}'}(\tau)$ and the reward-weighted previous one $p_{\boldsymbol{\theta}}(\tau) R(\tau)$. The Kullback-Leibler divergence is considered a natural distance measure between probability distributions (Bagnell and Schneider, 2003; van der Maaten et al, 2009). As shown in (Dayan and Hinton, 1997; Peters and Schaal, 2007), such a derivation results in a lower bound on the expected return using Jensen's inequality and the concavity of the logarithm. Thus, we obtain

$$\log J(\boldsymbol{\theta}') = \log \int_{\mathbb{T}} p_{\boldsymbol{\theta}'}(\tau) R(\tau) d\tau = \log \int_{\mathbb{T}} \frac{p_{\boldsymbol{\theta}}(\tau)}{p_{\boldsymbol{\theta}}(\tau)} p_{\boldsymbol{\theta}'}(\tau) R(\tau) d\tau,$$

$$\geq \int_{\mathbb{T}} p_{\boldsymbol{\theta}}(\tau) R(\tau) \log \frac{p_{\boldsymbol{\theta}'}(\tau)}{p_{\boldsymbol{\theta}}(\tau)} d\tau + \mathrm{const},$$

which is proportional to

$$-D\left(p_{\boldsymbol{\theta}}(\tau) R(\tau) \| p_{\boldsymbol{\theta}'}(\tau)\right) = L_{\boldsymbol{\theta}}(\boldsymbol{\theta}'),$$

where

$$D\left(p(\tau) \| q(\tau)\right) = \int p(\tau) \log \frac{p(\tau)}{q(\tau)} d\tau$$

denotes the Kullback-Leibler divergence, and the constant is needed for tightness of the bound. Note that $p_{\boldsymbol{\theta}}(\tau) R(\tau)$ is an improper probability distribution as pointed out by Dayan and Hinton (1997). The policy improvement step is equivalent to maximizing the lower bound on the expected return $L_{\boldsymbol{\theta}}(\boldsymbol{\theta}')$, and we will now show how it relates to previous policy learning methods.

Algorithm 4.1 'Vanilla' Policy Gradients (VPG)

Input: initial policy parameters $\boldsymbol{\theta}_0$
repeat

    Sample: Perform $h = \{1, \ldots, H\}$ rollouts using $\mathbf{a} = \boldsymbol{\theta}^{\mathrm{T}}\boldsymbol{\phi}(\mathbf{s}, t) + \boldsymbol{\varepsilon}_t$ with $[\varepsilon_t^n] \sim \mathcal{N}(0, (\sigma^{h,n})^2)$ as stochastic policy and collect all $(t, \mathbf{s}_t^h, \mathbf{a}_t^h, \mathbf{s}_{t+1}^h, \boldsymbol{\varepsilon}_t^h, r_{t+1}^h)$ for $t = \{1, 2, \ldots, T + 1\}$.

    Compute: Return $R^h = \sum_{t=1}^{T+1} r_t^h$, eligibility

$$\psi^{h,n} = \frac{\partial \log p\left(\boldsymbol{\tau}^h\right)}{\partial \theta^n} = \sum_{t=1}^{T} \frac{\partial \log \pi\left(a_t^h | s_t^h, t\right)}{\partial \theta^n} = \sum_{t=1}^{T} \frac{\varepsilon_t^{h,n}}{(\sigma_h^n)^2} \phi^n \left(s_t^{h,n}, t\right)$$

and baseline

$$b^n = \frac{\sum_{h=1}^{H} \left(\psi^{h,n}\right)^2 R^h}{\sum_{h=1}^{H} \left(\psi^{h,n}\right)^2}$$

for each parameter $n = \{1, \ldots, N\}$ from rollouts.

    Compute Gradient:

$$g_{\mathrm{VP}}^n = E\left\{\frac{\partial \log p(\boldsymbol{\tau}^h)}{\partial \theta^n}\left(R(\boldsymbol{\tau}^h) - b^n\right)\right\} = \frac{1}{H}\sum_{h=1}^{H} \psi^{h,n}(R^h - b^n).$$

    Update policy using

$$\boldsymbol{\theta}_{k+1} = \boldsymbol{\theta}_k + \alpha \mathbf{g}_{\mathrm{VP}}.$$

until Convergence $\boldsymbol{\theta}_{k+1} \approx \boldsymbol{\theta}_k$.

Resulting Policy Updates

In this section, we will discuss three different policy updates, which are directly derived from the results of Section 4.2.2. First, we show that policy gradients (Williams, 1992; Sutton et al, 1999; Lawrence et al, 2003; Tedrake et al, 2004; Peters and Schaal, 2006) can be derived from the lower bound $L_\theta(\boldsymbol{\theta}')$, which is straightforward from a supervised learning perspective (Binder et al, 1997). Subsequently, we show that natural policy gradients (Bagnell and Schneider, 2003; Peters and Schaal, 2006) can be seen as an additional constraint regularizing the change in the path distribution resulting from a policy update when improving the policy incrementally. Finally, we will show how expectation-maximization (EM) algorithms for policy learning can be generated.

Policy Gradients

When differentiating the function $L_\theta(\boldsymbol{\theta}')$ that defines the lower bound on the expected return, we directly obtain

$$\partial_{\boldsymbol{\theta}'} L_\theta(\boldsymbol{\theta}') = \int_{\mathbb{T}} p_\theta(\boldsymbol{\tau}) R(\boldsymbol{\tau}) \partial_{\boldsymbol{\theta}'} \log p_{\boldsymbol{\theta}'}(\boldsymbol{\tau}) d\boldsymbol{\tau} = E\left\{\left(\sum_{t=1}^{T} \partial_{\boldsymbol{\theta}'} \log \pi(\mathbf{a}_t | \mathbf{s}_t, t)\right) R(\boldsymbol{\tau})\right\},$$

where

$$\partial_{\boldsymbol{\theta}'} \log p_{\boldsymbol{\theta}'}\left(\boldsymbol{\tau}\right) = \sum_{t=1}^{T} \partial_{\boldsymbol{\theta}'} \log \pi\left(\mathbf{a}_t | \mathbf{s}_t, t\right)$$

denotes the log-derivative of the path distribution. As this log-derivative depends only on the policy we can estimate a gradient from rollouts, without having a model, by simply replacing the expectation by a sum. When $\boldsymbol{\theta}'$ is close to $\boldsymbol{\theta}$, we have the policy gradient estimator, which is widely known as episodic REINFORCE (Williams, 1992)

$$\lim_{\boldsymbol{\theta}' \to \boldsymbol{\theta}} \partial_{\boldsymbol{\theta}'} L_{\boldsymbol{\theta}}(\boldsymbol{\theta}') = \partial_{\boldsymbol{\theta}} J(\boldsymbol{\theta}).$$

See Algorithm 4.1 for an example implementation of this algorithm and Appendix 4.A.1 for the detailed steps of the derivation. A MATLAB implementation of this algorithm is available at `http://www.robot-learning.de/Member/JensKober`.

A reward, which precedes an action in a rollout, can neither be caused by the action nor cause an action in the same rollout. Thus, when inserting Equation (4.1) into Equation (4.2.2), all cross-products between r_t and $\partial_{\boldsymbol{\theta}'} \log \pi(\mathbf{a}_{t+\delta t} | \mathbf{s}_{t+\delta t}, t + \delta t)$ for $\delta t > 0$ become zero in expectation (Peters and Schaal, 2006). Therefore, we can omit these terms and rewrite the estimator as

$$\partial_{\boldsymbol{\theta}'} L_{\boldsymbol{\theta}}\left(\boldsymbol{\theta}'\right) = E\left\{\sum_{t=1}^{T} \partial_{\boldsymbol{\theta}'} \log \pi\left(\mathbf{a}_t | \mathbf{s}_t, t\right) Q^{\pi}(\mathbf{s}, \mathbf{a}, t)\right\},$$

where

$$Q^{\pi}\left(\mathbf{s}, \mathbf{a}, t\right) = E\left\{\sum_{\tilde{t}=t}^{T} r\left(\mathbf{s}_{\tilde{t}}, \mathbf{a}_{\tilde{t}}, \mathbf{s}_{\tilde{t}+1}, \tilde{t}\right) | \mathbf{s}_t = \mathbf{s}, \mathbf{a}_t = \mathbf{a}\right\}$$

is called the state-action value function (Sutton and Barto, 1998). Equation (4.2.2) is equivalent to the policy gradient theorem (Sutton et al, 1999) for $\boldsymbol{\theta}' \to \boldsymbol{\theta}$ in the infinite horizon case, where the dependence on time t can be dropped.

The derivation results in the episodic Natural Actor Critic as discussed in (Peters et al, 2003, 2005) when adding an additional cost in Equation (4.2.2) to penalize large steps away from the observed path distribution. Such a regularization can be achieved by restricting the amount of change in the path distribution and subsequently, determining the steepest descent for a fixed step away from the observed trajectories. Change in probability distributions is naturally measured using the Kullback-Leibler divergence, thus after adding the additional constraint of

$$D\left(p_{\boldsymbol{\theta}}\left(\boldsymbol{\tau}\right) \| p_{\boldsymbol{\theta}'}\left(\boldsymbol{\tau}\right)\right) \approx 0.5\left(\boldsymbol{\theta}' - \boldsymbol{\theta}\right)^{\mathrm{T}} \mathbf{F}\left(\boldsymbol{\theta}\right)\left(\boldsymbol{\theta}' - \boldsymbol{\theta}\right) = \delta$$

using a second-order expansion as an approximation where $\mathbf{F}(\boldsymbol{\theta})$ denotes the Fisher information matrix (Bagnell and Schneider, 2003; Peters et al, 2003, 2005). See Algorithm 4.2 for an example implementation of the episodic Natural Actor Critic. A MATLAB implementation of this algorithm is available at `http://www.robot-learning.de/Member/JensKober`.

Algorithm 4.2 episodic Natural Actor Critic (eNAC)

Input: initial policy parameters $\boldsymbol{\theta}_0$
repeat

Sample: Perform $h = \{1, \ldots, H\}$ rollouts using $\mathbf{a} = \boldsymbol{\theta}^{\mathrm{T}}\boldsymbol{\phi}(\mathbf{s}, t) + \boldsymbol{\varepsilon}_t$ with $[\varepsilon_t^n] \sim \mathcal{N}(0, (\sigma^{h,n})^2)$ as stochastic policy and collect all $(t, \mathbf{s}_t^h, \mathbf{a}_t^h, \mathbf{s}_{t+1}^h, \boldsymbol{\varepsilon}_t^h, r_{t+1}^h)$ for $t = \{1, 2, \ldots, T+1\}$.

Compute: Return $R^h = \sum_{t=1}^{T+1} r_t^h$ and eligibility $\psi^{h,n} = \sum_{t=1}^{T}(\sigma_h^n)^{-2}\varepsilon_t^{h,n}\phi^n(s_t^{h,n}, t)$ for each parameter $n = \{1, \ldots, N\}$ from rollouts.

Compute Gradient:

$$\left[\mathbf{g}_{\mathrm{eNAC}}^{\mathrm{T}}, R_{\mathrm{ref}}\right]^{\mathrm{T}} = \left(\boldsymbol{\Psi}^{\mathrm{T}}\boldsymbol{\Psi}\right)^{-1}\boldsymbol{\Psi}^{\mathrm{T}}\mathbf{R}$$

with $\mathbf{R} = \left[R^1, \ldots, R^H\right]^{\mathrm{T}}$ and $\boldsymbol{\Psi} = \begin{bmatrix} \psi^1, \ldots, \psi^H \\ 1, \ldots, 1 \end{bmatrix}^{\mathrm{T}}$ where $\psi^h = \left[\psi^{h,1}, \ldots, \psi^{h,N}\right]^{\mathrm{T}}$.

Update policy using

$$\boldsymbol{\theta}_{k+1} = \boldsymbol{\theta}_k + \alpha \mathbf{g}_{\mathrm{eNAC}}.$$

until Convergence $\boldsymbol{\theta}_{k+1} \approx \boldsymbol{\theta}_k$.

Policy Search via Expectation Maximization

One major drawback of gradient-based approaches is the learning rate, which is an open parameter that can be hard to tune in control problems but is essential for good performance. Expectation-Maximization algorithms are well-known to avoid this problem in supervised learning while even yielding faster convergence (McLachan and Krishnan, 1997). Previously, similar ideas have been explored in immediate reinforcement learning (Dayan and Hinton, 1997; Peters and Schaal, 2007). In general, an EM-algorithm chooses the next policy parameters $\boldsymbol{\theta}_{n+1}$ such that

$$\boldsymbol{\theta}_{n+1} = \mathrm{argmax}_{\boldsymbol{\theta}'}\, L_{\boldsymbol{\theta}}\left(\boldsymbol{\theta}'\right).$$

In the case where $\pi(\mathbf{a}_t|\mathbf{s}_t, t)$ belongs to the exponential family, the next policy can be determined analytically by setting Equation (4.2.2) or Equation (4.2.2) to zero

$$E\left\{\sum_{t=1}^{T}\partial_{\boldsymbol{\theta}'}\log\pi\left(\mathbf{a}_t|\mathbf{s}_t, t\right)Q\left(\mathbf{s}, \mathbf{a}, t\right)\right\} = 0,$$

and solving for $\boldsymbol{\theta}'$. Depending on the choice of stochastic policy, we will obtain different solutions and different learning algorithms. It allows the extension of the reward-weighted regression to longer horizons as well as the introduction of the Policy learning by Weighting Exploration with the Returns (PoWER) algorithm.

Algorithm 4.3 episodic Reward Weighted Regression (eRWR)

Input: initial policy parameters $\boldsymbol{\theta}_0$
repeat

Sample: Perform $h = \{1, \ldots, H\}$ rollouts using $\mathbf{a} = \boldsymbol{\theta}^{\mathrm{T}} \phi(\mathbf{s}, t) + \boldsymbol{\varepsilon}_t$ with $[\varepsilon_t^n] \sim \mathcal{N}(0, (\sigma^{h,n})^2)$ as stochastic policy and collect all $(t, \mathbf{s}_t^h, \mathbf{a}_t^h, \mathbf{s}_{t+1}^h, \boldsymbol{\varepsilon}_t^h, r_{t+1}^h)$ for $t = \{1, 2, \ldots, T+1\}$.

Compute: State-action value function $Q_t^{\pi,h} = \sum_{\tilde{t}=t}^{T} r_{\tilde{t}}^h$ from rollouts.

Update policy using

$$\boldsymbol{\theta}_{k+1}^n = \left((\boldsymbol{\Phi}^n)^{\mathrm{T}} \mathbf{Q}^{\pi} \boldsymbol{\Phi}^n \right)^{-1} (\boldsymbol{\Phi}^n)^{\mathrm{T}} \mathbf{Q}^{\pi} \mathbf{A}^n$$

with basis functions

$$\boldsymbol{\Phi}^n = \left[\phi_1^{1,n}, \ldots, \phi_T^{1,n}, \phi_1^{2,n}, \ldots, \phi_1^{H,n}, \ldots, \phi_T^{H,n} \right]^{\mathrm{T}},$$

where $\phi_t^{h,n}$ is the value of the basis function of rollout h and parameter n at time t,
actions

$$\mathbf{A}^n = \left[a_1^{1,n}, \ldots, a_T^{1,n}, a_1^{2,n}, \ldots, a_1^{H,n}, \ldots, a_T^{H,n} \right]^{\mathrm{T}},$$

and returns

$$\mathbf{Q}^{\pi} = \mathrm{diag}\left(Q_1^{\pi,1}, \ldots, Q_T^{\pi,1}, Q_1^{\pi,2}, \ldots, Q_1^{\pi,H}, \ldots, Q_T^{\pi,H} \right)$$

until Convergence $\boldsymbol{\theta}_{k+1} \approx \boldsymbol{\theta}_k$.

4.2.3 Policy learning by Weighting Exploration with the Returns (PoWER)

In most learning control problems, we attempt to have a deterministic mean policy $\bar{\mathbf{a}} = \boldsymbol{\theta}^{\mathrm{T}} \phi(\mathbf{s}, t)$ with parameters $\boldsymbol{\theta}$ and basis functions ϕ. In Section 4.3.1, we will introduce a particular type of basis function well-suited for robotics. These basis functions derive from the motor primitive formulation. Given such a deterministic mean policy $\bar{\mathbf{a}} = \boldsymbol{\theta}^{\mathrm{T}} \phi(\mathbf{s}, t)$, we generate a stochastic policy using additive exploration $\boldsymbol{\varepsilon}(\mathbf{s}, t)$ in order to make model-free reinforcement learning possible. We have a policy $\pi(\mathbf{a}_t | \mathbf{s}_t, t)$ that can be brought into the form

$$\mathbf{a} = \boldsymbol{\theta}^{\mathrm{T}} \phi(\mathbf{s}, t) + \boldsymbol{\epsilon}(\phi(\mathbf{s}, t)).$$

Previous work in this setting (Williams, 1992; Guenter et al, 2007; Peters and Schaal, 2006, 2007), with the notable exception of (Rückstieß et al, 2008), has focused on state-independent, white Gaussian exploration, namely $\boldsymbol{\epsilon}(\phi(\mathbf{s}, t)) \sim \mathcal{N}(\boldsymbol{\epsilon}|\mathbf{0}, \boldsymbol{\Sigma})$. It is straightforward to obtain the Reward-Weighted Regression for episodic RL by solving Equation (4.2.2) for $\boldsymbol{\theta}'$, which naturally yields a weighted regression method with the state-action values $Q^{\pi}(\mathbf{s}, \mathbf{a}, t)$ as weights. See Algorithm 4.3 for an exemplary implementation and Appendix 4.A.2 for the derivation. An optimized MATLAB implementation of this algorithm is available at `http://www.robot-learning.de/Member/JensKober`. This form of exploration has resulted in various applications in robotics such as T-Ball batting (Peters and Schaal, 2006), Peg-In-Hole (Gullapalli et al, 1994), constrained reaching movements (Guenter et al, 2007) and operational space control (Peters and Schaal, 2007).

However, such unstructured exploration at every step has several disadvantages: (i) it causes a large variance in parameter updates that grows with the number of time-steps (Rückstieß et al, 2008; Peters and Schaal, 2006), (ii) it perturbs actions too frequently as the system acts as a low pass filter, and the perturbations average out, thus their effect is washed out, and (iii) it can damage the system executing the trajectory. As the action is perturbed in every time-step the outcome of a trial can change drastically. This effect accumulates with the number of trials and the exploration is not equal over the progress of the trial. This behavior leads to a large variance in parameter updates. Random exploration in every time-step leads to jumps in the actions. A physical robot can not execute instantaneous changes in actions as either the controller needs time to react or the motor and the links of the robot have inertia that forces the robot to continue the motion induced by the previous actions. Globally speaking, the system acts as a low pass filter. If the robot tries to follow the desired high frequency action changes, a lot of strain is placed on the mechanics of the robot and can lead to oscillations. Furthermore, the accumulating effect of the exploration can lead the robot far from previously seen states, which is potentially dangerous.

As a result, all methods relying on this state-independent exploration have proven too fragile for learning tasks such as the Ball-in-a-Cup (see Section 4.3.7) on a real robot system. Alternatively, as introduced in (Rückstieß et al, 2008), one could generate a form of structured, state-dependent exploration. We use

$$\epsilon \left(\phi \left(\mathbf{s}, t \right) \right) = \varepsilon_t^{\mathrm{T}} \phi \left(\mathbf{s}, t \right)$$

with $\varepsilon_t \sim \mathcal{N}(\mathbf{0}, \hat{\Sigma})$, where $\hat{\Sigma}$ is a hyper-parameter of the exploration that can be optimized in a similar manner (see Appendix 4.A.3). This argument results in the policy

$$\mathbf{a} \sim \pi \left(\mathbf{a}_t | \mathbf{s}_t, t \right) = \mathcal{N} \left(\mathbf{a} | \boldsymbol{\theta}^{\mathrm{T}} \phi \left(\mathbf{s}, t \right), \phi(\mathbf{s}, t)^{\mathrm{T}} \hat{\Sigma} \phi(\mathbf{s}, t) \right).$$

Inserting the resulting policy into Equation (4.2.2), we obtain the optimality condition update and can derive the update rule

$$\boldsymbol{\theta}' = \boldsymbol{\theta} + E \left\{ \sum_{t=1}^{T} \mathbf{W} \left(\mathbf{s}, t \right) Q^{\pi} \left(\mathbf{s}, \mathbf{a}, t \right) \right\}^{-1} E \left\{ \sum_{t=1}^{T} \mathbf{W} \left(\mathbf{s}, t \right) \varepsilon_t Q^{\pi} \left(\mathbf{s}, \mathbf{a}, t \right) \right\}$$

with $\mathbf{W}(\mathbf{s}, t) = \phi(\mathbf{s}, t)\phi(\mathbf{s}, t)^{\mathrm{T}}(\phi(\mathbf{s}, t)^{\mathrm{T}} \hat{\Sigma} \phi(\mathbf{s}, t))^{-1}$.

In order to reduce the number of rollouts in this on-policy scenario, we reuse the rollouts through importance sampling as described, in the context of reinforcement learning, in (Andrieu et al, 2003; Sutton and Barto,

---

Algorithm 4.4 EM Policy learning by Weighting Exploration with the Returns (PoWER)

Input: initial policy parameters $\boldsymbol{\theta}_0$

repeat

    Sample: Perform   rollout(s)   using $\mathbf{a} = (\boldsymbol{\theta} + \boldsymbol{\varepsilon}_t)^{\mathrm{T}}\boldsymbol{\phi}(\mathbf{s},t)$ with $\boldsymbol{\varepsilon}_t^{\mathrm{T}}\boldsymbol{\phi}(\mathbf{s},t) \sim \mathcal{N}(0,\boldsymbol{\phi}(\mathbf{s},t)^{\mathrm{T}}\hat{\boldsymbol{\Sigma}}\boldsymbol{\phi}(\mathbf{s},t))$ as   stochastic   policy   and   collect   all $(t,\mathbf{s}_t,\mathbf{a}_t,\mathbf{s}_{t+1},\boldsymbol{\varepsilon}_t,r_{t+1})$ for $t = \{1,2,\dots,T+1\}$.

    Estimate: Use unbiased estimate

$$\hat{Q}^{\pi}(\mathbf{s},\mathbf{a},t) = \sum_{\tilde{t}=t}^{T} r\left(\mathbf{s}_{\tilde{t}},\mathbf{a}_{\tilde{t}},\mathbf{s}_{\tilde{t}+1},\tilde{t}\right).$$

    Reweight: Compute importance weights and reweight rollouts, discard low-importance rollouts.

    Update policy using

$$\boldsymbol{\theta}_{k+1} = \boldsymbol{\theta}_k + \left\langle \sum_{t=1}^{T} \mathbf{W}(\mathbf{s},t)\, Q^{\pi}(\mathbf{s},\mathbf{a},t) \right\rangle_{w(\boldsymbol{\tau})}^{-1} \left\langle \sum_{t=1}^{T} \mathbf{W}(\mathbf{s},t)\, \boldsymbol{\varepsilon}_t Q^{\pi}(\mathbf{s},\mathbf{a},t) \right\rangle_{w(\boldsymbol{\tau})}$$

    with $\mathbf{W}(\mathbf{s},t) = \boldsymbol{\phi}(\mathbf{s},t)\boldsymbol{\phi}(\mathbf{s},t)^{\mathrm{T}}(\boldsymbol{\phi}(\mathbf{s},t)^{\mathrm{T}}\hat{\boldsymbol{\Sigma}}\boldsymbol{\phi}(\mathbf{s},t))^{-1}$.

until Convergence $\boldsymbol{\theta}_{k+1} \approx \boldsymbol{\theta}_k$.

1998). The expectations $E\{\cdot\}$ are replaced by the importance sampler denoted by $\langle\cdot\rangle_{w(\boldsymbol{\tau})}$. To avoid the fragility sometimes resulting from importance sampling in reinforcement learning, samples with very small importance weights are discarded. This step is necessary as a lot of rollouts with a low return accumulate mass and can bias the update. A simple heuristic that works well in practice is to discard all but the j best rollouts, where j is chosen in the same order of magnitude as the number of parameters N. The derivation is shown in Appendix 4.A.3 and the resulting algorithm in Algorithm 4.4. Note that for our motor primitives, some simplifications of $\mathbf{W}$ are possible. These and other simplifications are shown in Appendix 4.A.3. A MATLAB implementation of this algorithm in several variants is available at `http://www.robot-learning.de/Member/JensKober`. As we will see in Section 4.3, this PoWER method significantly outperforms all other described methods.

PoWER is very robust with respect to reward functions. The key constraint is that it has to be an improper probability distribution which means that the rewards have to be positive. It can be beneficial for learning speed if the reward function sums up to one as a proper probability distribution.

Like most learning algorithms, PoWER achieves convergence faster for lower numbers of parameters. However, as it is an EM-inspired approach, it suffers significantly less from this problem than gradient based approaches. Including more prior knowledge, either in the parametrization or the initial policy, leads to faster convergence. As discussed above, changing exploration at every time-step has a number of disadvantages. Fixing the exploration for the whole episode (if each basis function is only active for a short time) or using a slowly varying exploration (for example based on random walks) can increase the performance. All algorithms introduced in this chapter optimize locally and can get stuck in local optima. An initial policy should be chosen to avoid local optima on the progress towards the desired final solution.

4.3 Benchmark Evaluation and Application in Robotics

In this section, we demonstrate the effectiveness of the algorithms presented in Section 4.2.3 in the context of motor primitive learning for robotics. We will first give a quick overview of how the motor primitives (Ijspeert et al, 2002b,a; Schaal et al, 2003, 2007) work and how learning algorithms can be used to adapt them. Subsequently, we will discuss how we can turn the parametrized motor primitives (Ijspeert et al, 2002b,a; Schaal et al, 2003, 2007) into explorative, stochastic policies (Rückstieß et al, 2008). We show that the novel PoWER algorithm outperforms many previous well-known methods, particularly 'Vanilla' Policy Gradients (Williams, 1992; Sutton et al, 1999; Lawrence et al, 2003; Peters and Schaal, 2006), Finite Difference Gradients (Sehnke et al, 2010; Peters and Schaal, 2006), the episodic Natural Actor Critic (Peters et al, 2003, 2005), and the generalized Reward-Weighted Regression (Peters and Schaal, 2007) on the two simulated benchmark problems suggested by Peters and Schaal (2006) and the Underactuated Swing-Up (Atkeson, 1994). We compare policy search based algorithms to a value function based one on the mountain-car benchmark. Additionally, we evaluate policy search methods on the multidimensional robotic tasks Tetherball Target Hitting and Casting. As a significantly more complex motor learning task, we will show how the robot can learn a high-speed Ball-in-a-Cup movement (Sumners, 1997) with motor primitives for all seven degrees of freedom of our Barrett WAM robot arm. An overview of the experiments is presented in Table 4.1.

4.3.1 Dynamical Systems Motor Primitives as Stochastic Policies

In the analytically tractable cases, episodic Reinforcement Learning (RL) problems have been studied deeply in the optimal control community. In this field it is well-known that for a finite horizon problem, the optimal solution is non-stationary (Kirk, 1970) and, in general, cannot be represented by a time-independent policy. The motor primitives based on dynamical systems (Ijspeert et al, 2002b,a; Schaal et al, 2003, 2007) represent a particular type of time-variant policy that has an internal phase, which corresponds to a clock with additional flexibility (for example, for

Table 4.1 Overview of the Experiments: 4.3.2 Basic Motor Learning, 4.3.3 Mountain-Car, 4.3.4 Tetherball Target Hitting, 4.3.5 Underactuated Swing-Up, 4.3.6 Casting, and 4.3.7 Ball-in-a-Cup

	Open Parameters	*DoF*	*Rollouts*	*Policy*	*Platform*	*Algorithms*
4.3.2	10 (shape)	1	4400	MP	simulation	FDG, VPG, eNAC, eRWR, PoWER
4.3.3	2 (switching)	1	80	bang-bang	simulation	FDG, PoWER, kNN-TD(λ)
4.3.4	6 (positions)	1	200	rhythmic	simulation	FDG, PoWER
4.3.5	10 (goal & shape)	1	200/100	MP	simu/robot	FDG, VPG, eNAC, eRWR, PoWER
4.3.6	10 (shape)	2	200	MP	simulation	eNAC, PoWER
4.3.7	217 (shape)	7	100	MP	robot	PoWER

incorporating coupling effects, perceptual influences, etc.). Thus, they can represent optimal solutions for finite horizons. We embed this internal clock or movement phase into our state and from an optimal control perspective have ensured that the optimal solution can be represented. See Chapter 3 for a more detailed discussion.

One of the biggest advantages of this motor primitive framework (Ijspeert et al, 2002b,a; Schaal et al, 2003, 2007) is that the second system, in Equation (3.1), is linear in the policy parameters $\boldsymbol{\theta}$ and is therefore well-suited for both imitation learning as well as for the presented reinforcement learning algorithms. For example, if we would have to learn only a motor primitive for a single degree of freedom q_i, then we could use a motor primitive in the form $\ddot{q}_i = \boldsymbol{\phi}(\mathbf{s})^{\mathrm{T}}\boldsymbol{\theta}$ where $\mathbf{s} = [q_i, \dot{q}_i, z]$ is the state and where time is implicitly embedded in z. We use the output of $\ddot{\bar{q}}_i = \boldsymbol{\phi}(\mathbf{s})^{\mathrm{T}}\boldsymbol{\theta} = \bar{a}$ as the policy mean. The perturbed accelerations $\ddot{q}_i = a = \bar{a} + \varepsilon$ are given to the system.

In Sections 4.3.5 and 4.3.7, we use imitation learning from a single example to generate a sensible initial policy. This step can be performed efficiently in the context of dynamical systems motor primitives as the policy is linear in its parameters, see Section 3.2.4.

4.3.2   Benchmark Comparison I: Basic Motor Learning Examples

As a benchmark comparison, we follow a previously studied scenario in order to evaluate, which method is best-suited for our problem class. We perform our evaluations on exactly the same benchmark problems as in (Peters and Schaal, 2006) and use two tasks commonly studied in motor control literature for which the analytic solutions are known. The first task is a reaching task, wherein a goal has to be reached at a certain time, while the used

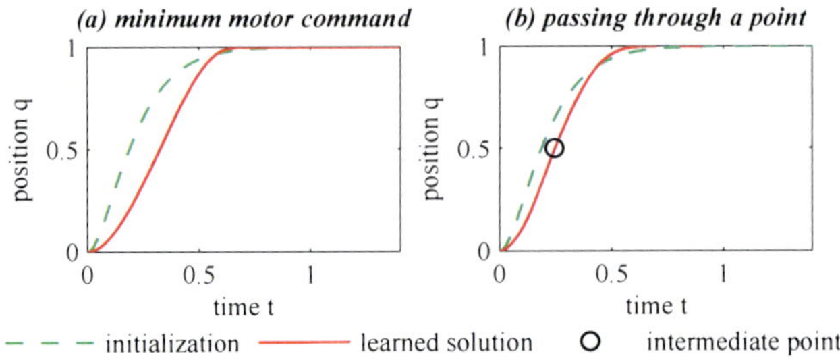

Fig. 4.1 This figure shows the initial and the final trajectories for the two motor control tasks. Both start at 0 and have to go to 1 with minimal accelerations. From $T/2 = 0.75$ on the trajectory has to be as close to 1 as possible. For the passing through task the trajectory additionally has to pass through $p_M = 0.5$ at time $M = 7/40T$ indicated by the circle.

motor commands have to be minimized. The second task is a reaching task of the same style with an additional via-point. The task is illustrated in Figure 4.1. This comparison mainly shows the suitability of our algorithm (Algorithm 4.4) and that it outperforms previous methods such as Finite Difference Gradient (FDG) methods (Sehnke et al, 2010; Peters and Schaal, 2006), see Algorithm 4.5, 'Vanilla' Policy Gradients (VPG) with optimal baselines (Williams, 1992; Sutton et al, 1999; Lawrence et al, 2003; Peters and Schaal, 2006), see Algorithm 4.1, the episodic Natural Actor Critic (eNAC) (Peters et al, 2003, 2005), see Algorithm 4.2, and the new episodic version of the Reward-Weighted Regression (eRWR) algorithm (Peters and Schaal, 2007), see Algorithm 4.3. MATLAB implementations of all algorithms are available at `http://www.robot-learning.de/Member/JensKober`. For all algorithms except PoWER, we used batches to update the policy. A sliding-window based approach is also possible. For VPG, eNAC, and eRWR a batch size of $H = 2N$ and for FDG a batch size of $H = N+1$ are typical. For PoWER, we employed an importance sampling based approach, although a batch based update is also possible.

We consider two standard tasks taken from (Peters and Schaal, 2006), but we use the newer form of the motor primitives from (Schaal et al, 2007). The first task is to achieve a goal with a minimum-squared movement acceleration and a given movement duration, that gives a return of

$$R\left(\boldsymbol{\tau}\right) = -\sum_{t=0}^{T/2} c_1 \ddot{q}_t^2 - \sum_{t=T/2+1}^{T} c_2 \left(\left(q_t - g\right)^2 + \dot{q}_t^2\right)$$

for optimization, where $T = 1.5$, $c_1 = 1/100$ is the weight of the transient rewards for the movement duration $T/2$, while $c_2 = 1000$ is the importance of the final reward, extended over the time interval $[T/2 + 1, T]$ which insures that the goal state $g = 1.0$ is reached and maintained properly. The initial state of the motor primitive is always zero in this evaluation.

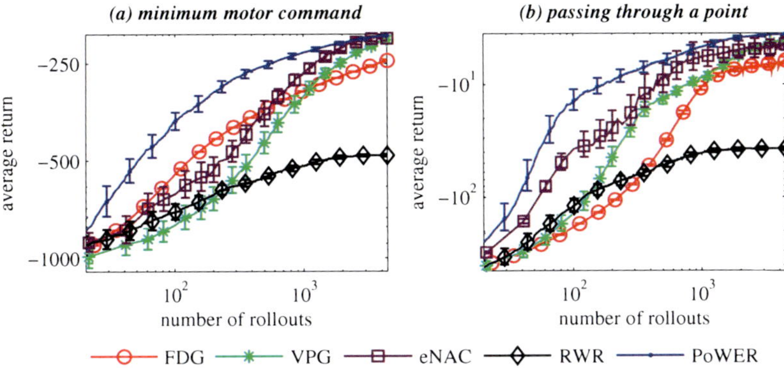

Fig. 4.2 This figure shows the mean performance of all compared methods in two benchmark tasks averaged over twenty learning runs with the error bars indicating the standard deviation. Policy learning by Weighting Exploration with the Returns (PoWER) clearly outperforms Finite Difference Gradients (FDG), 'Vanilla' Policy Gradients (VPG), the episodic Natural Actor Critic (eNAC), and the adapted Reward-Weighted Regression (eRWR) for both tasks. Note that this plot has logarithmic scales on both axes, thus a unit difference corresponds to an order of magnitude. The omission of the first twenty rollouts was necessary to cope with the log-log presentation.

The second task involves passing through an intermediate point during the trajectory, while minimizing the squared accelerations, that is, we have a similar return with an additional punishment term for missing the intermediate point p_M at time M given by

$$R\left(\boldsymbol{\tau}\right) = -\sum_{t=0}^{T/2}\widetilde{c}_1\ddot{q}_t^2 - \sum_{t=T/2+1}^{T}\widetilde{c}_2\left(\left(q_t - g\right)^2 + \dot{q}_t^2\right) - \widetilde{c}_3\left(q_M - p_M\right)^2$$

where $\widetilde{c}_1 = 1/10000$, $\widetilde{c}_2 = 200$, $\widetilde{c}_3 = 20000$. The goal is given by $g = 1.0$, the intermediate point a value of $p_M = 0.5$ at time $M = 7/40T$, and the initial state was zero. This return yields a smooth movement, which passes through the intermediate point before reaching the goal, even though the optimal solution is not identical to the analytic solution with hard constraints.

All open parameters were manually optimized for each algorithm in order to maximize the performance while not destabilizing the convergence of the learning process. When applied in the episodic scenario, Policy learning by Weighting Exploration with the Returns (PoWER) clearly outperformed the episodic Natural Actor Critic (eNAC), 'Vanilla' Policy Gradient (VPG), Finite Difference Gradient (FDG), and the episodic Reward-Weighted Regression (eRWR) for both tasks. The episodic Reward-Weighted Regression (eRWR) is outperformed by all other algorithms suggesting that this algorithm does not generalize well from the immediate reward case. While FDG gets stuck on a plateau, both eNAC and VPG converge to the same good

---

Algorithm 4.5 Finite Difference Gradients (FDG)

Input: initial policy parameters $\boldsymbol{\theta}_0$

repeat

Generate policy variations: $\Delta\boldsymbol{\theta}^h \sim \mathcal{U}_{[-\Delta\boldsymbol{\theta}_{\min},\Delta\boldsymbol{\theta}_{\max}]}$ for $h = \{1,\ldots,H\}$ rollouts.

Sample: Perform $h = \{1,\ldots,H\}$ rollouts using $\mathbf{a} = (\boldsymbol{\theta}+\Delta\boldsymbol{\theta}^h)^{\mathrm{T}}\boldsymbol{\phi}(\mathbf{s},t)$ as policy and collect all $(t,\mathbf{s}_t^h,\mathbf{a}_t^h,\mathbf{s}_{t+1}^h,\boldsymbol{\varepsilon}_t^h,r_{t+1}^h)$ for $t = \{1,2,\ldots,T+1\}$.

Compute: Return $R^h(\boldsymbol{\theta}+\Delta\boldsymbol{\theta}^h) = \sum_{t=1}^{T+1} r_t^h$ from rollouts.

Compute Gradient:

$$\left[\mathbf{g}_{\mathrm{FD}}^{\mathrm{T}}, R_{\mathrm{ref}}\right]^{\mathrm{T}} = \left(\Delta\boldsymbol{\Theta}^{\mathrm{T}}\Delta\boldsymbol{\Theta}\right)^{-1}\Delta\boldsymbol{\Theta}^{\mathrm{T}}\mathbf{R}$$

with $\Delta\boldsymbol{\Theta} = \begin{bmatrix}\Delta\boldsymbol{\theta}^1,\ldots,\Delta\boldsymbol{\theta}^H \\ 1,\ldots,1\end{bmatrix}^{\mathrm{T}}$ and $\mathbf{R} = \left[R^1,\ldots,R^H\right]^{\mathrm{T}}$.

Update policy using

$$\boldsymbol{\theta}_{k+1} = \boldsymbol{\theta}_k + \alpha\mathbf{g}_{\mathrm{FD}}.$$

until Convergence $\boldsymbol{\theta}_{k+1} \approx \boldsymbol{\theta}_k$.

---

final solution. PoWER finds the a slightly better solution while converging noticeably faster. The results are presented in Figure 4.2.

4.3.3   Benchmark Comparison II: Mountain-Car

As a typical reinforcement learning benchmark we chose the mountain-car task (Sutton and Barto, 1998) as it can be treated with episodic reinforcement learning. In this problem we have a car placed in a valley, and it is supposed to go on the top of the mountain in front of it, but does not have the necessary capabilities of acceleration to do so directly. Thus, the car has to first drive up the mountain on the opposite side of the valley to gain sufficient energy. The dynamics are given in (Sutton and Barto, 1998) as

$$\dot{x}_{t+1} = \dot{x}_t + 0.001a_t - 0.0025\cos\left(3x_t\right),$$

$$x_{t+1} = x_t + \dot{x}_{t+1},$$

with position $-1.2 \leq x_{t+1} \leq 0.5$ and velocity $-0.07 \leq \dot{x}_{t+1} \leq 0.07$. If the goal $x_{t+1} \geq 0.5$ is reached the episode is terminated. If the left bound is reached the velocity is reset to zero. The initial condition of the car is $x_0 = -0.5$ and $\dot{x}_0 = 0$. The reward is $r_t = -1$ for all time-steps until the car reaches the goal. We employed an undiscounted return. The set of actions a_t is slightly different to the setup proposed by Sutton and Barto (1998). We only have two actions, the full throttle forward ($a_t = +1$) and the full throttle reverse ($a_t = -1$). From a classical optimal control point of view, it

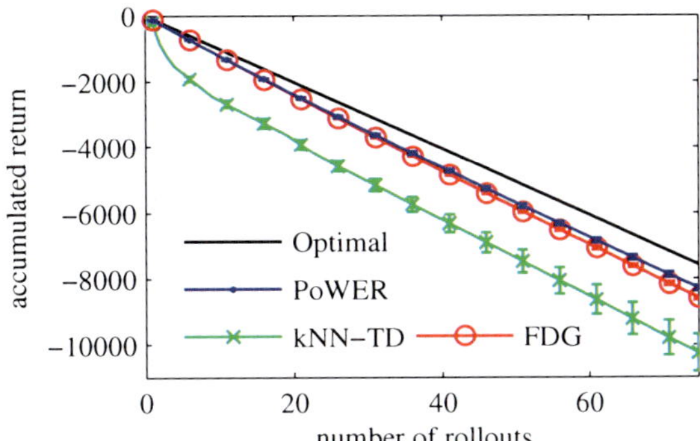

(a) The tasks consists of driving the underpowered car to the target on the mountain indicated by the yellow star.

(b) This figure shows the mean accumulated returns of the methods compared on the mountain car benchmark. The results are averaged over fifty learning runs with error bars inidicating the standard deviation. Policy learning by Weighting Exploration with the Returns (PoWER) and Finite Difference Gradients (FDG) clearly outperform kNN-TD(λ). All methods converge to the optimal solution.

Fig. 4.3 This figure shows an illustration of the mountain-car task and the mean accumulated returns of the compared methods

is straightforward to see that a bang-bang controller can solve this problem. As an initial policy we chose a policy that accelerates forward until the car cannot climb the mountain further, accelerates reverse until the car cannot climb the opposite slope further, and finally accelerates forward until the car reaches the goal. This policy reaches the goal but is not optimal as the car can still accumulate enough energy if it reverses the direction slightly earlier. As a parametrization for the policy search approaches we chose to encode the switching points of the acceleration. The two parameters of the policy indicate at which timestep t the acceleration is reversed. For this kind of policy only algorithms that perturb the parameters are applicable and we compare a Finite Difference Gradient approach to PoWER. This parametrized policy is entirely different to motor primitives. Additionally we included a comparison to a value function based method. The Q-function was initialized with our initial policy. As the kNN-TD(λ) algorithm (Martín H. et al, 2009) won the Reinforcement Learning Competitions in 2008 and 2009, we selected it for this comparison. This comparison is contrived as our switching policy always starts in a similar initial state while the value function based policy can start in a wider range of states. Furthermore, the policy search approaches may be sped up by the initialization, while kNN-TD(λ) will learn the optimal policy without prior knowledge and does not benefit much from the initialization. However, the use of a global approach, such as kNN-TD(λ) requires a global

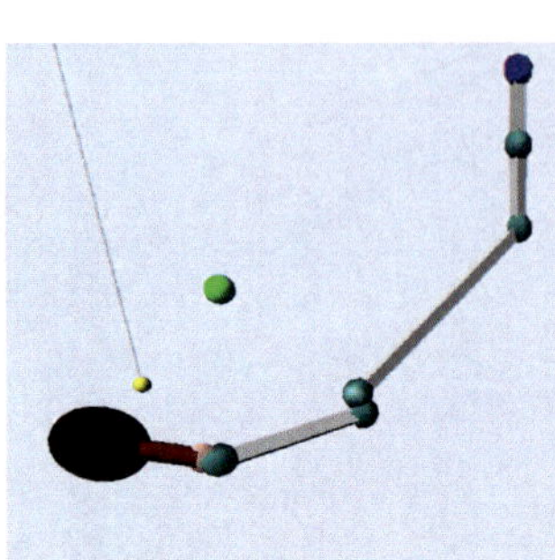

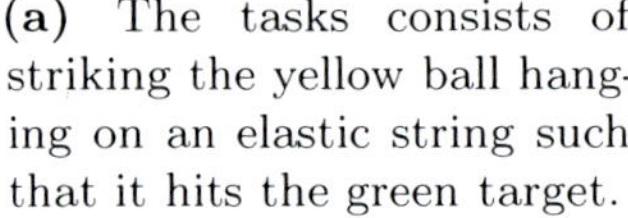

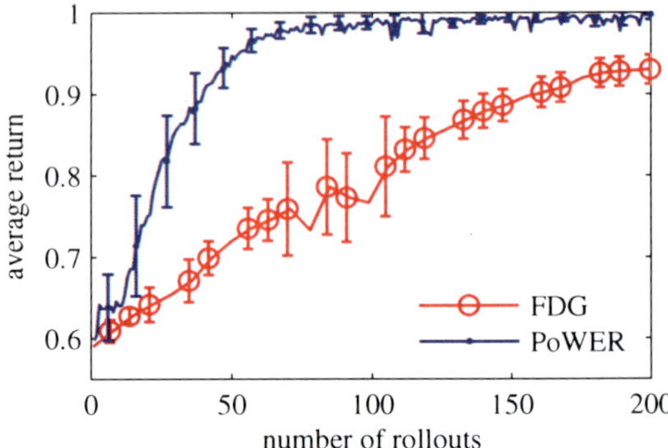

(a) The tasks consists of striking the yellow ball hanging on an elastic string such that it hits the green target.

(b) The returns are averaged over 20 learning runs with error bars indicating the standard deviation. Policy learning by Weighting Exploration with the Returns (PoWER) clearly outperforms Finite Difference Gradients (FDG).

Fig. 4.4 This figure shows an illustration of the Tetherball Target Hitting task and the mean returns of the compared methods

search of the state space. Such a global search limits the scalability of these approaches. The more local approaches of policy search are less affected by these scalability problems. Figure 4.3b shows the performance of these algorithms. As kNN-TD(λ) initially explores the unseen parts of the Q-function, the policy search approaches converge faster. All methods find the optimal solution.

4.3.4 Benchmark Comparison III: Tetherball Target Hitting

In this task, a table tennis ball is hanging on an elastic string from the ceiling. The task consists of hitting the ball with a table tennis racket so that it hits a fixed target. The task is illustrated in Figure 4.4a. The return is based on the minimum distance between the ball and the target during one episode transformed by an exponential. The policy is parametrized as the position of the six lower degrees of freedom of the Barrett WAM. Only the first degree of freedom (shoulder rotation) is moved during an episode. The movement is represented by a rhythmic policy with a fixed amplitude and period. Due to the parametrization of the task only PoWER and Finite Difference Gradients are applicable. We observed reliable performance if the initial policy did not miss the target by more than approximately $50cm$. In this experiment it took significantly less iterations to find a good initial policy than to tune the learning rate of Finite Difference Gradients, a problem from which PoWER

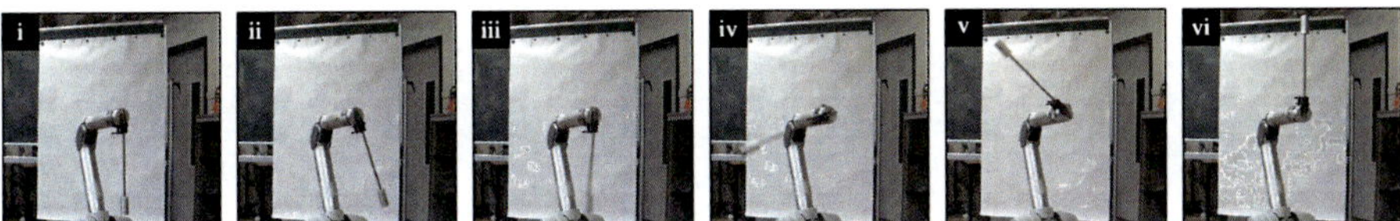

Fig. 4.5 This figure shows the time series of the Underactuated Swing-Up where only a single joint of the robot is moved with a torque limit ensured by limiting the maximal motor current of that joint. The resulting motion requires the robot to (ii) first move away from the target to limit the maximal required torque during the swing-up in (iii-v) and subsequent stabilization (vi).

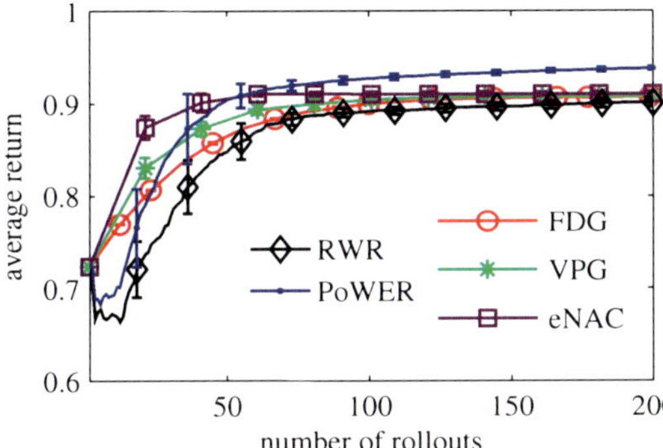

Fig. 4.6 This figure shows the performance of all compared methods for the swing-up in simulation and the mean performance averaged over 20 learning runs with the error bars indicating the standard deviation. PoWER outperforms the other algorithms from 50 rollouts on and finds a significantly better policy.

did not suffer as it is an EM-like algorithm. Figure 4.4b illustrates the results. PoWER converges significantly faster.

4.3.5   Benchmark Comparison IV: Underactuated Swing-Up

As an additional simulated benchmark and for the real-robot evaluations, we employed the Underactuated Swing-Up (Atkeson, 1994). Here, only a single degree of freedom is represented by the motor primitive as described in Section 4.3.1. The goal is to move a hanging heavy pendulum to an upright position and to stabilize it there. The objective is threefold: the pendulum has to be swung up in a minimal amount of time, has to be stabilized in the upright position, and achieve these goals with minimal motor torques. By limiting the motor current for this degree of freedom, we can ensure that the torque limits described in (Atkeson, 1994) are maintained and directly moving the joint to the right position is not possible. Under these torque limits, the robot needs to first move away from the target to reduce the maximal required torque during the swing-up, see Figure 4.5. This problem is similar to the mountain-car problem (Section 4.3.3). The standard mountain-car problem is designed to get the car to the top of the mountain in minimum time. It does

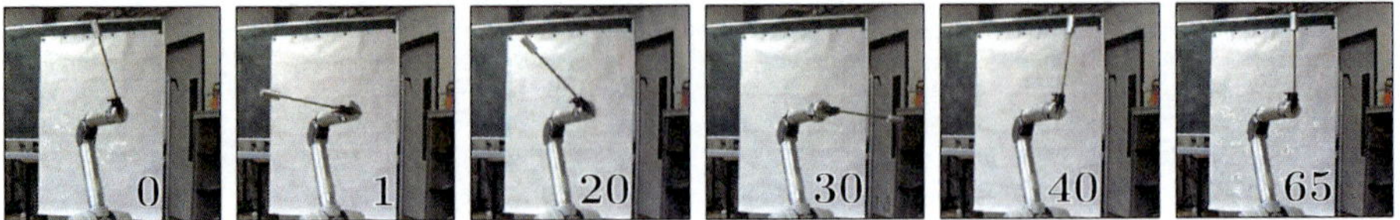

Fig. 4.7 This figure shows the improvement of the policy over rollouts. The snapshots from the video show the final positions. (0) Initial policy after imitation learning (without torque limit). (1) Initial policy after imitation learning (with active torque limit). (20) Policy after 20 rollouts, going further up. (30) Policy after 30 rollouts, going too far. (40) Policy after 40 rollouts, going only a bit too far. (65) Final policy after 65 rollouts.

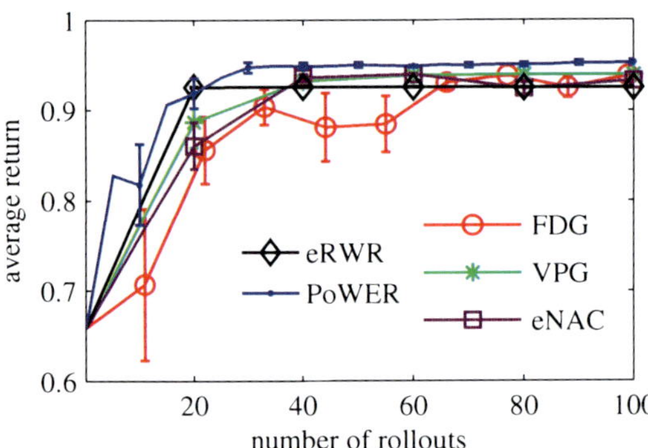

Fig. 4.8 This figure shows the performance of all compared methods for the swing-up on the real robot and the mean performance averaged over 3 learning runs with the error bars indicating the standard deviation. PoWER outperforms the other algorithms and finds a significantly better policy.

not matter if it stops at this point or drives at a high speed as usage of the accelerator and brake is not punished. Adding the requirement of stabilizing the car at the top of the mountain makes the problem significantly harder. These additional constraints exist in the Underactuated Swing-Up task where it is required that the pendulum (the equivalent of the car) stops at the top to fulfill the task. The applied torque limits were the same as in (Atkeson, 1994) and so was the reward function, except that the complete return of the trajectory was transformed by an $\exp(\cdot)$ to ensure positivity. The reward function is given by

$$r\left(t\right) = -c_1 q\left(t\right)^2 + c_2 \log\cos\left(c_3 \frac{u\left(t\right)}{u_{\max}}\right),$$

where the constants are $c_1 = 5/\pi^2 \approx 0.507$, $c_2 = (2/\pi)^2 \approx 0.405$, and $c_3 = \pi/2 \approx 1.571$. Please note that π refers to the mathematics constant here, and not to the policy. The first term of the sum is punishing the distance to the desired upright position $q = 0$, and the second term is punishing the usage of motor torques u. A different trade-off can be achieved by changing the parameters or the structure of the reward function, as long as it remains an improper probability function. Again all open parameters of all algorithms were manually optimized. The motor primitive with nine shape parameters and one goal parameter was initialized by imitation learning from a kinesthetic teach-in. Kinesthetic teach-in means "taking the robot by the hand", performing the task by moving the robot while it is in gravity-compensation mode, and recording the joint angles, velocities, and accelerations. This initial demonstration needs to include all the relevant features of the movement, e.g., it should first move away from the target and then towards the upright position. The performance of the algorithms is very robust, as long as the initial policy with active torque limits moves the pendulum approximately above the horizontal orientation.

As the batch size, and, thus the learning speed, of the gradient based approaches depend on the number of parameters (see Section 4.3.2), we tried to minimize the number of parameters. Using more parameters would allow us to control more details of the policy which could result in a better final policy, but would have significantly slowed down convergence. At least nine shape parameters were needed to ensure that the imitation can capture the movement away from the target, which is essential to accomplish the task. We compared all algorithms considered in Section 4.3.2 and could show that PoWER would again outperform the others. The convergence is a lot faster than in the basic motor learning examples (see Section 4.3.2), as we do not start from scratch but rather from an initial solution that allows significant improvements in each step for each algorithm. The results are presented in Figure 4.6. See Figure 4.7 and Figure 4.8 for the resulting real-robot performance.

4.3.6   Benchmark Comparison V: Casting

In this task a ball is attached to the endeffector of the Barrett WAM by a string. The task is to place the ball into a small cup in front of the robot. The task is illustrated in Figure 4.9a. The return is based on the sum of the minimum distance between the ball and the top, the center, and the bottom of the cup respectively during one episode. Using only a single distance, the return could be successfully optimized, but the final behavior often corresponded to a local maximum (for example hitting the cup from the side). The movement is in a plane and only one shoulder DoF and the elbow are moved. The policy is parametrized using motor primitives with five shape parameters per active degree of freedom. The policy is initialized with a movement that results in hitting the cup from the side in the upper quarter of the cup. If the ball hits the cup below the middle, approximately 300 rollouts were required for PoWER and we did not achieve reliable performance for the episodic Natural Actor Critic. We compare the two best performing algorithms from the basic motor learning examples (see Section 4.3.2) and the Underactuated Swing-Up (see Section 4.3.5), namely eNAC and PoWER. Figure 4.9b illustrates the results. PoWER again converges significantly faster.

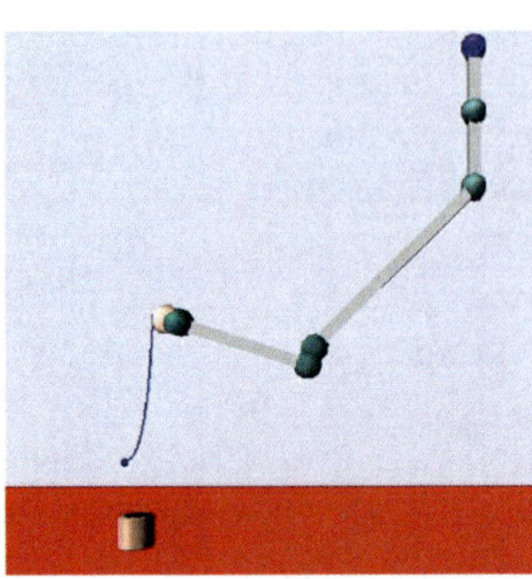

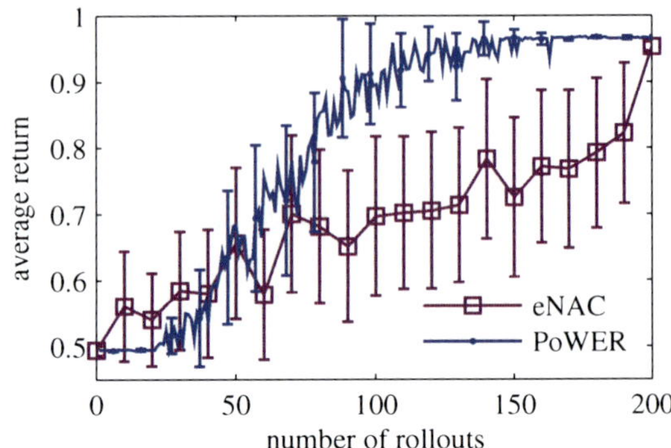

(a) The tasks consists of placing the blue ball in the brown cup.

(b) The returns are averaged over 20 learning runs with error bars indicating the standard deviation. Policy learning by Weighting Exploration with the Returns (PoWER) clearly outperforms episodic Natural Actor Critic (eNAC).

Fig. 4.9 This figure illustrates the Casting task and shows the mean returns of the compared methods

4.3.7 Ball-in-a-Cup on a Barrett WAM

The children's motor skill game Ball-in-a-Cup (Sumners, 1997), also known as Balero, Bilboquet, and Kendama, is challenging even for adults. The toy has a small cup which is held in one hand (or, in our case, is attached to the end-effector of the robot) and the cup has a small ball hanging down on a string (the string has a length of $40cm$ in our setup). Initially, the ball is hanging down vertically in a rest position. The player needs to move fast in order to induce a motion in the ball through the string, toss it up, and catch it with the cup. A possible movement is illustrated in Figure 4.11 in the top row.

Note that learning of Ball-in-a-Cup and Kendama has previously been studied in robotics. We are going to contrast a few of the approaches here. While we learn directly in the joint space of the robot, Takenaka (1984) recorded planar human cup movements and determined the required joint movements for a planar, three degree of freedom (DoF) robot, so that it could follow the trajectories while visual feedback was used for error compensation. Both Sato et al (1993) and Shone et al (2000) used motion planning approaches which relied on very accurate models of the ball and the string while employing only one DoF in (Shone et al, 2000) or two DoF in (Sato et al, 1993) so that the complete state-space could be searched exhaustively. Interestingly, exploratory robot moves were used in (Sato et al, 1993) to estimate the parameters of the employed model. Probably the most advanced preceding work on learning Kendama was carried out by Miyamoto et al

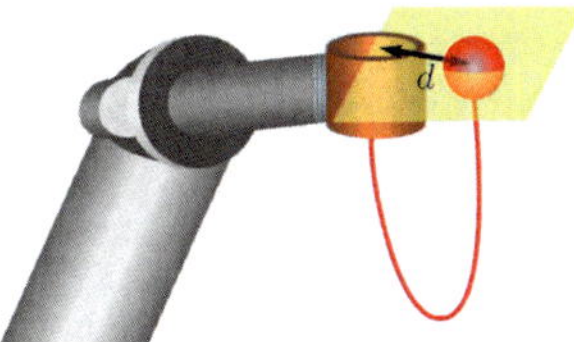

Fig. 4.10 This figure illustrates how the reward is calculated. The plane represents the level of the upper rim of the cup. For a successful rollout the ball has to be moved above the cup first and is then flying in a downward direction into the opening of the cup. The reward is calculated as the distance d of the center of the cup and the center of the ball on the plane at the moment the ball is passing the plane in a downward direction. If the ball is flying directly into the center of the cup, the distance is 0 and through the transformation $\exp(-d^2)$ yields the highest possible reward of 1. The further the ball passes the plane from the cup, the larger the distance and thus the smaller the resulting reward.

(1996) who used a seven DoF anthropomorphic arm and recorded human motions to train a neural network to reconstruct via-points. Employing full kinematic knowledge, the authors optimize a desired trajectory.

The state of the system can be described by joint angles and joint velocities of the robot as well as the the Cartesian coordinates and velocities of the ball. The actions are the joint space accelerations where each of the seven joints is driven by a separate motor primitive, but with one common canonical system. The movement uses all seven degrees of freedom and is not in a plane. All motor primitives are perturbed separately but employ the same joint final reward given by

$$
r(t) = \begin{cases} \exp\left(-\alpha\,(x_c - x_b)^2 - \alpha\,(y_c - y_b)^2\right) & \text{if } t = t_c, \\ 0 & \text{otherwise,} \end{cases}
$$

where we denote the moment when the ball passes the rim of the cup with a downward direction by t_c, the cup position by $[x_c, y_c, z_c] \in \mathbb{R}^3$, the ball position by $[x_b, y_b, z_b] \in \mathbb{R}^3$, and a scaling parameter by $\alpha = 100$ (see also Figure 4.10). The algorithm is robust to changes of this parameter as long as the reward clearly discriminates good and suboptimal trials. The directional information is necessary as the algorithm could otherwise learn to hit the bottom of the cup with the ball. This solution would correspond to a local maximum whose reward is very close to the optimal one, but the policy very far from the optimal one. The reward needs to include a term avoiding this local maximum. PoWER is based on the idea of considering the reward as an improper probability distribution. Transforming the reward using the exponential enforces this constraint. The reward is not only affected by the movements of the cup but foremost by the movements of the ball, which are sensitive to small changes in the cup's movement. A small perturbation of the initial condition or during the trajectory can change the movement of the

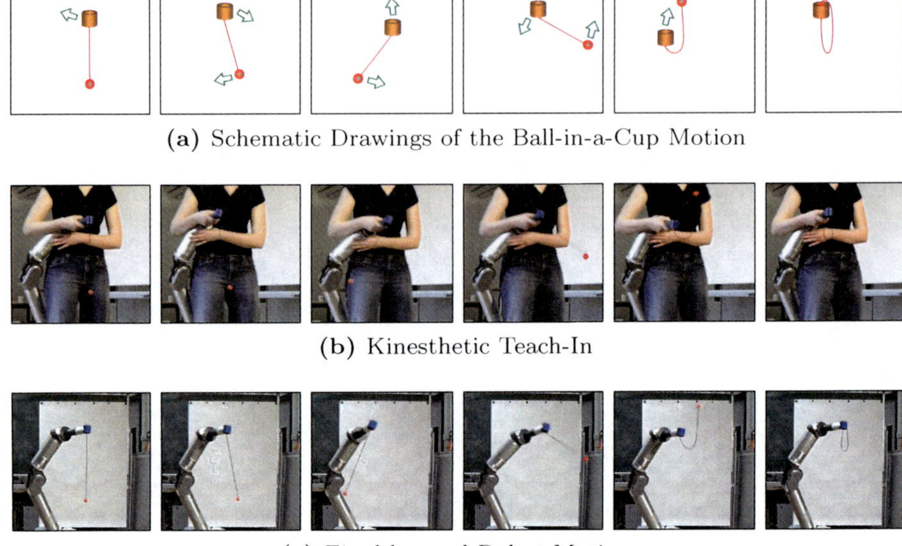

(a) Schematic Drawings of the Ball-in-a-Cup Motion

(b) Kinesthetic Teach-In

(c) Final learned Robot Motion

Fig. 4.11 This figure shows schematic drawings of the Ball-in-a-Cup motion (a), the final learned robot motion (c), as well as a kinesthetic teach-in (b). The arrows show the directions of the current movements in that frame. The human cup motion was taught to the robot by imitation learning with 31 parameters per joint for an approximately 3 seconds long trajectory. The robot manages to reproduce the imitated motion quite accurately, but the ball misses the cup by several centimeters. After approximately 75 iterations of our Policy learning by Weighting Exploration with the Returns (PoWER) algorithm the robot has improved its motion so that the ball goes in the cup. Also see Figure 4.12.

ball and hence the outcome of the complete movement. The position of the ball is estimated using a stereo vision system and is needed to determine the reward.

Due to the complexity of the task, Ball-in-a-Cup is a hard motor learning task for children, who usually only succeed at it by observing another person playing combined with a lot of improvement by trial-and-error. Mimicking how children learn to play Ball-in-a-Cup, we first initialize the motor primitives by imitation learning and, subsequently, improve them by reinforcement learning. We recorded the motions of a human player by kinesthetic teach-in to obtain an example for imitation as shown in Figure 4.11b. A single demonstration was used for imitation learning. Learning from multiple demonstrations did not improve the performance as the task is sensitive to small differences. As expected, the robot fails to reproduce the presented behavior even if we use all the recorded details for the imitation. Thus, reinforcement learning is needed for self-improvement. The more parameters used

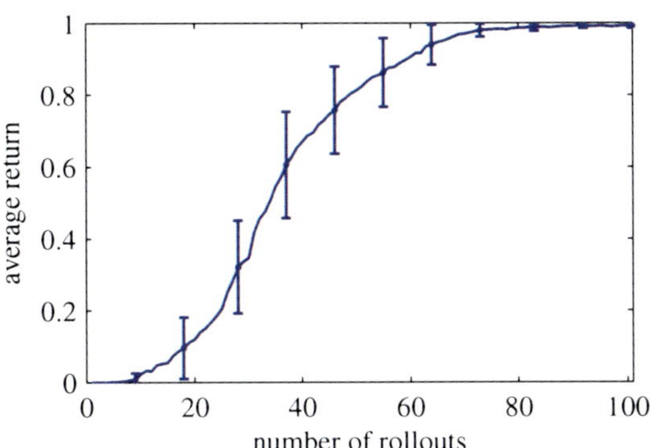

Fig. 4.12 This figure shows the expected return of the learned policy in the Ball-in-a-Cup evaluation averaged over 20 runs

for the learning, the slower the convergence is. Due to the imitation, the ball must go above the rim of the cup such that the algorithm gets at least a small positive reward for all rollouts. This way exhaustive exploration is avoided as the algorithm can compare the performance of the different rollouts. We determined that 31 shape-parameters per motor primitive are needed. With less parameters the ball does not go above the rim of the cup in the initial trial and the algorithm does not receive any meaningful information about the trial using the aforementioned reward function. More shape-parameters will lead to a more accurate reproduction of the demonstrated movement and, thus, to a better initial policy. However, there is a trade-off between this better initial policy and a potentially lower learning speed. Using three times as many parameters the algorithm converged at roughly the same time. The hyper-parameters σ_{ij} are initially set in the same order of magnitude as the median of the parameters for each motor primitive and are then optimized alongside the shape-parameters by PoWER. The performance of the algorithm is fairly robust for values chosen in this range. Figure 4.12 shows the expected return over the number of rollouts where convergence to a maximum is clearly recognizable. The robot regularly succeeds at bringing the ball into the cup after approximately 75 iterations. Figure 4.13 shows the improvement of the policy over the rollouts. From our experience, nine year old children get the ball in the cup for the first time after about 35 trials while the robot gets the ball in for the first time after 42-45 rollouts. However, after 100 trials, the robot exhibits perfect runs in every single trial while children do not have a comparable success rate. Of course, such a comparison with children is contrived as a robot can precisely reproduce movements unlike any human being, and children can most likely adapt faster to changes in the setup.

Fig. 4.13 This figure shows the improvement of the policy over rollouts. The snapshots from the video show the position of the ball closest to the cup during a rollout. (1) Initial policy after imitation learning. (15) Policy after 15 rollouts, already closer. (25) Policy after 25 rollouts, going too far. (45) Policy after 45 rollouts, hitting the near rim. (60) Policy after 60 rollouts, hitting the far rim. (100) Final policy after 100 rollouts.

4.4   Discussion and Conclusion

In Section 4.4.1, we will discuss robotics as a benchmark for reinforcement learning, in Section 4.4.2 we discuss different simulation to robot transfer scenarios, and we will draw our conclusions in Section 4.4.3.

4.4.1   Discussion: Robotics as Benchmark for Reinforcement Learning?

Most reinforcement learning algorithms are evaluated on synthetic benchmarks, often involving discrete states and actions. Agents in many simple grid worlds take millions of actions and episodes before convergence. As a result, many methods have focused too strongly on such problem domains. In contrast, many real world problems such as robotics are best represented with high-dimensional, continuous states and actions. Every single trial run is costly and as a result such applications force us to focus on problems that will often be overlooked accidentally in synthetic examples. Simulations are a helpful testbed for debugging algorithms. Continuous states and actions as well as noise can be simulated, however simulations pose the temptation of using unrealistically many iterations and also allow us to exploit the perfect models.

Our typical approach consists of testing the algorithm in a simulation of the robot and the environment. Once the performance is satisfactory we replace either the robot or the environment with its real counterpart depending on the potential hazards. Replacing the robot is always possible as we can still simulate the environment taking into account the state of the real robot. Learning with a simulated robot in the real environment is not always a possibility especially if the robot influences the observed environment, such as in the Ball-in-a-Cup task. The final evaluations are done on the real robot in the real environment.

Our experience in robotics show that the plant and the environment can often not be represented accurately enough by a physical model and that the learned policies are thus not entirely transferable. If sufficiently accurate models were available, we could resort to the large body of work on optimal control (Kirk, 1970), which offers alternatives to data driven reinforcement learning. However, when a model with large errors is used, the solution suffers severely from an optimization bias as already experienced by Atkeson (1994). Here, the reinforcement learning algorithm exploits the imperfections of the simulator rather than yielding an optimal policy.

None of our learned policies could be transferred from a simulator to the real system without changes despite that the immediate errors of the simulator have been smaller than the measurement noise. Methods which jointly learn the models and the policy as well as perform some of the evaluations and updates in simulation (such as Dyna-like architectures as in (Sutton, 1990)) may alleviate this problem. In theory, a simulation could also be used to eliminate obviously bad solutions quickly. However, the differences between the simulation and the real robot do accumulate over time and this approach is only feasible if the horizon is short or the simulation very accurate. Priming the learning process by imitation learning and optimizing in its vicinity achieves the desired effect better and has thus been employed in this chapter.

Parametrized policies greatly reduce the need of samples and evaluations. Choosing an appropriate representation like motor primitives renders the learning process even more efficient.

One major problem with robotics benchmarks is the repeatability of the experiments. The results are tied to the specific hardware and setup used. Even a comparison with simulated robots is often not possible as many groups rely on proprietary simulators that are not freely available. Most of the benchmarks presented in this chapter rely on initial demonstrations to speed up the learning process. These demonstrations play an important part in the learning process. However, these initial demonstrations are also tied to the hardware or the simulator and are, thus, not straightforward to share. Comparing new algorithms to results from different authors usually requires the reimplementation of their algorithms to have a fair comparison.

Reproducibility is a key requirement for benchmarking but also a major challenge for robot reinforcement learning. To overcome this problem there are two orthogonal approaches: (i) a central organizer provides an identical setup for all participants and (ii) all participants share the same setup in a benchmarking lab. The first approach has been majorly pushed by funding agencies in the USA and Europe. In the USA, there have been special programs on robot learning such as DARPA Learning Locomotion (L2), Learning Applied to Ground Robotics (LAGR) and the DARPA Autonomous Robot Manipulation (ARM) (DARPA, 2010c,b,a). However, the hurdles involved in getting these robots to work have limited the participation to strong robotics groups instead of opening the robot reinforcement learning domain to more machine learning researchers. Alternative ways of providing identical setups are low cost standardized hardware or a system composed purely of commercially available components. The first suffers from reproducibility and reliability issues while the latter results in significant system integration problems. Hence, it may be more suitable for a robot reinforcement learning challenge to be hosted by a robot learning group with significant experience in both domains. The host lab specifies tasks that they have been able to accomplish on a real robot system. The hosts also need to devise a virtual robot laboratory for allowing the challenge participants to program, test and debug their algorithms. To limit the workload and the risk of the organizers, a first benchmarking round would be conducted using this simulated setup to determine the promising approaches. Successful participants will be invited by the host lab in order to test these algorithms in learning on the real system where the host lab needs to provide significant guidance. To our knowledge, no benchmark based on this approach has been proposed yet. We are currently evaluating possibilities to organize a challenge using such a shared setup in the context of the PASCAL2 Challenge Program (PASCAL2, 2010).

To successfully apply reinforcement learning to robotics, a fair level of knowledge on the limitations and maintenance of the robot hardware is necessary. These limitations include feasible actions, feasible run-time, as well as measurement delay and noise. Cooperation with a strong robotics group is still extremely important in order to apply novel reinforcement learning methods in robotics.

4.4.2   Discussion on Simulation to Robot Transfer Scenarios

In this chapter, we have discussed reinforcement learning for real robots with highly dynamic tasks. The opposite extreme in robotics would be, for example, a maze navigation problem where a mobile robot that has macro-actions such as "go left" and the lower level control moves the robot exactly by a well-defined distance to the left. In this scenario, it would probably be easier to transfer simulation results to real systems. For highly dynamic tasks or environments, accurate simulations are generically difficult. Besides fast moving objects and many interacting objects as well as deformable objects (often called soft bodies), like cloth, string, fluids, hydraulic tubes and other elastic materials are hard to simulate reliably and, thus, have an enormous impact on transferability. Additionally, the level and quality of measurement noise has a direct implication on the difficulty and the transferability of the learning task.

Better simulations often alleviate some of these problems. However, there is always a trade-off as more detailed simulations also require more precise model identification, higher temporal resolution, and, frequently even finite elements based simulations. Such detailed simulations may even be much slower than real-time, thus defeating one of the major advantages of simulations.

Aside from these clear difficulties in creating simulations that allow the transfer to real systems, we have observed three critically different scenarios for reinforcement learning in robotics. These scenarios are characterized by the energy flow between the policy and the system. In the energy-absorbing scenario, the task has passive dynamics and, hence, it is safer and easier to learn on a real robot. We are going to discuss the examples of Ball-Paddling, foothold selection in legged locomotion, and grasping (see Section 4.4.2). The second scenario has a border-line behavior: the system conserves most of the energy but the policy also only needs to inject energy into the system for a limited time. We will discuss Ball-in-a-Cup, Tetherball Target Hitting, and Mountain-Car as examples for this scenario (see Section 4.4.2). In the energy-emitting scenario energy is inserted due to the system dynamics even if the policy does not transfer energy into the system. The classical examples are Cart-Pole and inverted helicopters, and we also have the Underactuated Swing-Up which has to stabilize at the top (see Section 4.4.2). These different scenarios have implications on the relative utility of simulations and real robots.

As we are discussing our experience in performing such experiments, it may at times appear anecdotal. We hope the reader benefits from our insights nevertheless. However, the resulting classification bears similarities with insights on control law derivation (Fantoni and Lozano, 2001).

Energy-Absorbing Scenario

In this scenario, the system absorbs energy from the actions. As shown in Figure 4.14, we learned a Ball-Paddling task where a ping-pong ball is attached to a paddle by an elastic string and the ball has to be kept in the air by repeatedly hitting it from below. In this setup, the elastic string pulls the ball back towards the center of the paddle and the contact forces between the ball and the paddle are very complex. We modeled the system in as much detail as possible, including friction models, restitution models, dampening models, models for the elastic string, and air drag. However, in simulation the paddling behavior was still highly unpredictable and we needed a few thousand iterations to learn an optimal frequency, amplitude, and movement shape. The number of simulated trials exceeded the feasible amount on a real system. In contrast, when learning on the real system, we obtained a stable paddling behavior by imitation learning using the initial demonstration only and no further reinforcement learning was needed.

In general, scenarios with complex contact forces often work better in a real-world experiment. This problem was particularly drastic in locomotion experiments on rough terrain where the real world was an easier learning environment due to favorable friction properties during foot contact (Peters, 2007). In this experiment, learning was significantly harder in simulation and the learned policy could not be transferred. The same effect occurs in grasping when objects often cannot be grasped in simulation due to slip

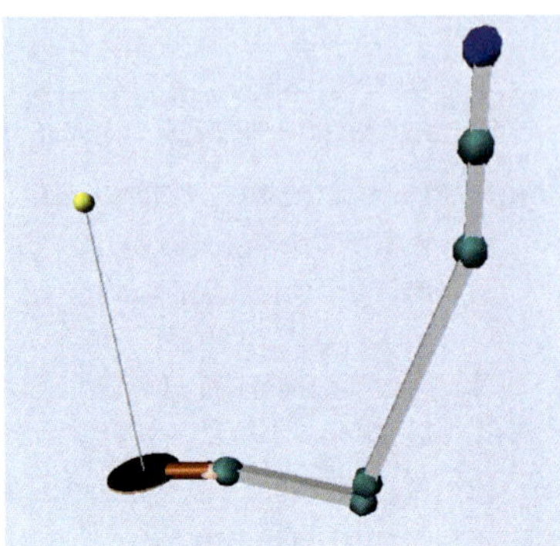

(a) Reinforcement learning required unrealistically many trials in simulation.

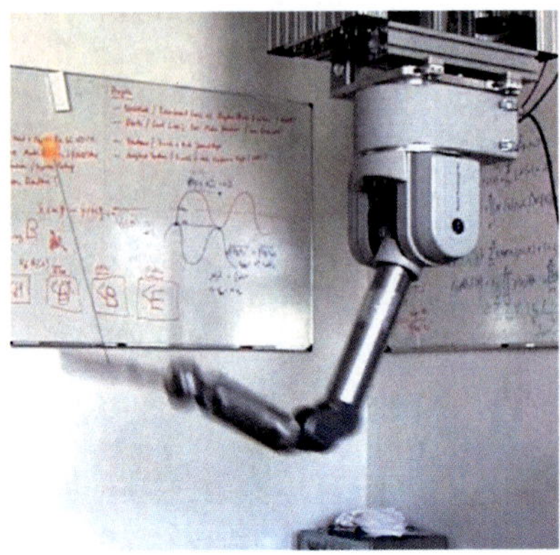

(b) Imitation learning only was sufficient on the real robot.

Fig. 4.14 This figure illustrates the Ball-Paddling task in simulation and on the real robot. The difference between simulation and robotics can be particularly emphasized in this problem where unrealistically many trials were needed on the simulation for reinforcement learning while the real world behavior could be learned by imitation learning. It illustrates the energy-consuming scenario and the difficulties of realistic learning in the presence of contact forces.

but the real world friction still allows them to be picked up. Hence, in this scenario, policies from simulations are frequently not helpful and learning in simulation is harder than on the real system. The results only transfer in a few cases. A simulation is therefore only recommended as a feasibility study and for software debugging. As most contacts differ significantly due to the current properties (which vary with time and temperature) of the two interacting objects, only a learned simulator is likely to grasp all relevant details.

Border-Line Scenario

In this scenario, adding too much energy to a system does not necessarily lead to a negative outcome. For example, in the Mountain-Car problem (see Section 4.3.3), inserting more energy and driving through the goal at a higher velocity does not affect task achievement. In contrast not inserting enough energy will result in a failure as the car cannot reach the top of the mountain. The Tetherball Target Hitting application presented in Section 4.3.4 exhibits a very similar behavior. The Ball-in-a-Cup experiment (see Section 4.3.7) highlights the resulting similarities between learning in good simulations and the real world for this scenario. Success is possible if more energy is inserted and the ball flies higher. However, when using too little energy the ball will stay below the opening of the cup. In this favorable scenario the "classical" strategy can be applied: learn how to learn in simulation. The policy learned in simulation does not necessarily transfer to the real world and the real-world scenario can be highly different but the learning speed and behavior are similar. Hence, hyper-parameters such as learning and exploration rates can be tuned in simulation. The learning algorithm may take longer due to increased errors, modeling problems and uncertainty. Still, good practice is to create a sufficiently accurate simulator and to adapt the learning strategy subsequently to the real system.

Energy-Emitting Scenario

Energy emission causes very different problems. Uncertainty in states will cause overreactions, hence, drastic failures are likely to occur when the system becomes unstable in a control theory sense. This system excitability often makes the task significantly harder to learn on a real robot in comparison to a simulated setup. Here, pre-studies in simulations are a necessary but not sufficient condition. Due to unmodeled nonlinearities, the exploration will affect various algorithms differently. Classical examples are helicopters in inverted flight (Ng et al, 2004b) and the pendulum in a Cart-Pole task in an upright position (Sutton and Barto, 1998) as these have to be constantly stabilized. Additionally the pendulum in the Swing-Up has to be stabilized in the final position or it will fall over and cause a failure. In this chapter, we take the example of the Swing-Up to illustrate how some methods unexpectedly do better in the real world as exhibited by Figures 4.6 and 4.8. The learning progress of all algorithms is noisier and the eRWR performs better on the real robot. The form of exploration employed by PoWER seems to give it an additional advantage in the first 20 rollouts as direct exploration on the actions is partially obscured by measurement noise. In order to cope with differences to the real-world, simulations need to be more stochastic than the real system (as suggested by Ng et al (2004b)) and should be learned to make transferring the results easier (as for example in (Schaal et al, 2002)).

4.4.3   Conclusion

In this chapter, we have presented a framework for deriving several policy learning methods that are applicable in robotics and an application to a highly complex motor learning task on a real Barrett WAM robot arm. We have shown that policy gradient methods are a special case of this framework. During initial experiments, we realized that the form of exploration highly influences the speed of the policy learning method. This empirical insight resulted in a novel policy learning algorithm, Policy learning by Weighting Exploration with the Returns (PoWER), an EM-inspired algorithm that outperforms several other policy search methods both on standard benchmarks as well as on a simulated Underactuated Swing-Up.

We have successfully applied this novel PoWER algorithm in the context of learning two tasks on a physical robot, namely the Underacted Swing-Up and Ball-in-a-Cup. Due to the curse of dimensionality, we cannot start with an arbitrary solution. Instead, we mimic the way children learn Ball-in-a-Cup and first present an example movement for imitation learning, which is recorded using kinesthetic teach-in. Subsequently, our reinforcement learning algorithm learns how to move the ball into the cup reliably. After only realistically few episodes, the task can be regularly fulfilled and the robot shows very good average performance. After 100 rollouts, the hyper-parameters, such as the exploration rate, have converged to negligible size and do not influence the outcome of the behavior any further. The experiments in this chapter use the original variant of the motor primitives which cannot deal with large perturbations. However, the extended variable-feedback variant presented in (Kober et al, 2008) can deal with a variety of changes directly (for example, in the length of the string or the size or weight of the ball) while the approach presented in this chapter will recover quickly by learning an adjusted policy in a few roll-outs. In (Kober et al, 2008), we have also shown that learning a strategy of pulling the ball up and moving the cup under the ball (as in Kendama) is possible in approximately the same number of trials. We have discovered a variety of different human strategies for Ball-in-a-Cup in movement recordings, see (Chiappa et al, 2009).

Our approach has already given rise to follow-up work in other contexts, for example, (Vlassis et al, 2009; Kormushev et al, 2010). Theodorou et al (2010) have shown that an algorithm very similar to PoWER can also be derived from a completely different perspective, that is, the path integral approach.

4.A   Derivations

In this appendix, we provide the derivations of various algorithms in more details than in the main text. We first present, how the episodic REINFORCE (Williams, 1992) can be obtained (Section 4.A.1). Subsequently, we show how the episodic Reward Weighted Regression (eRWR) (Peters and Schaal, 2007) can be generalized for the episodic case (Section 4.A.2), and finally we derive EM Policy learning by Weighting Exploration with the Returns (PoWER) and show a number of simplifications (Section 4.A.3).

4.A.1   Reinforce

If we choose a stochastic policy in the form $a = \boldsymbol{\theta}^{\mathrm{T}}\boldsymbol{\phi}\left(\mathbf{s},t\right) + \varepsilon_t$ with $\varepsilon_t \sim \mathcal{N}\left(0,\sigma^2\right)$, we have

$$\pi\left(a_t | \mathbf{s}_t, t\right) = \frac{1}{\sigma\sqrt{2\pi}}\exp\left(-\frac{1}{2\sigma^2}\left(a - \boldsymbol{\theta}^{\mathrm{T}}\boldsymbol{\phi}\right)^2\right),$$

and, thus,

$$\partial_{\boldsymbol{\theta}} \log \pi = \sigma^{-2} \left(a - \boldsymbol{\theta}^{\mathrm{T}} \boldsymbol{\phi}\right) \boldsymbol{\phi}^{\mathrm{T}}.$$

Therefore, the gradient, in Equation (4.2.2), becomes

$$\partial_{\boldsymbol{\theta}'} L_{\boldsymbol{\theta}}(\boldsymbol{\theta}') = E\left\{\textstyle\sum_{t=1}^{T}\sigma^{-2} \left(a - \boldsymbol{\theta}'^{\mathrm{T}}\boldsymbol{\phi}\right) \boldsymbol{\phi}^{\mathrm{T}} R\right\} = E\left\{\textstyle\sum_{t=1}^{T}\sigma^{-2}\varepsilon_t \boldsymbol{\phi}^{\mathrm{T}} R\right\},$$

which corresponds to the episodic REINFORCE algorithm (Williams, 1992).

4.A.2 *Episodic Reward Weighted Regression (eRWR)*

Setting Equation (4.A.1) to zero

$$\partial_{\boldsymbol{\theta}'} L_{\boldsymbol{\theta}}(\boldsymbol{\theta}') = E\left\{\textstyle\sum_{t=1}^{T}\sigma^{-2} \left(a - \boldsymbol{\theta}'^{\mathrm{T}}\boldsymbol{\phi}\right) \boldsymbol{\phi}^{\mathrm{T}} R\right\} = 0,$$

we obtain

$$E\left\{\textstyle\sum_{t=1}^{T}\sigma^{-2} a R \boldsymbol{\phi}^{\mathrm{T}}\right\} = E\left\{\textstyle\sum_{t=1}^{T}\sigma^{-2} \left(\boldsymbol{\theta}'^{\mathrm{T}}\boldsymbol{\phi}\right) R \boldsymbol{\phi}^{\mathrm{T}}\right\}.$$

Since σ is constant, we have $E\{\sum_{t=1}^{T} a R \boldsymbol{\phi}^{\mathrm{T}}\} = \boldsymbol{\theta}'^{\mathrm{T}} E\{\sum_{t=1}^{T} \boldsymbol{\phi} R \boldsymbol{\phi}^{\mathrm{T}}\}$. The $\boldsymbol{\theta}'$ minimizing the least squares error can be found by locally weighted linear regression (R as weights and $\boldsymbol{\phi}$ as basis functions) considering each time-step and rollout separately

$$\boldsymbol{\theta}' = \left(\boldsymbol{\Phi}^{\mathrm{T}}\mathbf{R}\boldsymbol{\Phi}\right)^{-1}\boldsymbol{\Phi}^{\mathrm{T}}\mathbf{R}\mathbf{A},$$

with $\boldsymbol{\Phi} = [\boldsymbol{\phi}_1^1, \ldots, \boldsymbol{\phi}_T^1, \boldsymbol{\phi}_1^2, \ldots, \boldsymbol{\phi}_1^H, \ldots, \boldsymbol{\phi}_T^H]^{\mathrm{T}}$, $\mathbf{R} = \mathrm{diag}(R^1, \ldots, R^1, R^2, \ldots, R^H, \ldots, R^H)$, and $\mathbf{A} = [a_1^1, \ldots, a_T^1, a_1^2, \ldots, a_1^H, \ldots, a_T^H]^{\mathrm{T}}$ for H rollouts.

The same derivation holds if we use Equation (4.2.2) instead of Equation (4.2.2). Then $\mathbf{R}$ in the regression is replaced by $\mathbf{Q}^{\pi} = \mathrm{diag}(Q_1^{\pi,1}, \ldots, Q_T^{\pi,1}, Q_1^{\pi,2}, \ldots, Q_1^{\pi,H}, \ldots, Q_T^{\pi,H})$. Using the state-action value function Q^{π} yields slightly faster convergence than using the return R.

4.A.3 *EM Policy Learning by Weighting Exploration with the Returns (PoWER)*

If we choose a stochastic policy in the form $a = \left(\boldsymbol{\theta} + \boldsymbol{\varepsilon}_t\right)^{\mathrm{T}} \boldsymbol{\phi}\left(\mathbf{s}, t\right)$ with $\boldsymbol{\varepsilon}_t \sim \mathcal{N}(\mathbf{0}, \hat{\boldsymbol{\Sigma}})$, we have

$$\pi\left(a_t | \mathbf{s}_t, t\right) = \mathcal{N}\left(\mathbf{a} | \boldsymbol{\theta}^{\mathrm{T}}\boldsymbol{\phi}\left(\mathbf{s}, t\right), \boldsymbol{\phi}(\mathbf{s}, t)^{\mathrm{T}}\hat{\boldsymbol{\Sigma}}\boldsymbol{\phi}(\mathbf{s}, t)\right)$$

$$= \left(2\pi\boldsymbol{\phi}^{\mathrm{T}}\hat{\boldsymbol{\Sigma}}\boldsymbol{\phi}\right)^{-1/2} \exp\left(\frac{-\left(a - \boldsymbol{\theta}^{\mathrm{T}}\boldsymbol{\phi}\right)^2}{2\boldsymbol{\phi}^{\mathrm{T}}\hat{\boldsymbol{\Sigma}}\boldsymbol{\phi}}\right),$$

and, thus, $\partial_{\boldsymbol{\theta}} \log \pi = \left(a - \boldsymbol{\theta}^{\mathrm{T}}\boldsymbol{\phi}\right)\boldsymbol{\phi}^{\mathrm{T}}/\left(\boldsymbol{\phi}^{\mathrm{T}}\hat{\boldsymbol{\Sigma}}\boldsymbol{\phi}\right)$. Therefore Equation (4.2.2) becomes

$$\partial_{\boldsymbol{\theta}'} L_{\boldsymbol{\theta}}(\boldsymbol{\theta}') = E\left\{\sum_{t=1}^{T} \frac{\left(a - \boldsymbol{\theta}'^{\mathrm{T}}\boldsymbol{\phi}\right)\boldsymbol{\phi}^{\mathrm{T}}}{\boldsymbol{\phi}^{\mathrm{T}}\hat{\boldsymbol{\Sigma}}\boldsymbol{\phi}} Q^{\pi}\right\}.$$

Setting this equation to zero is equivalent to

$$E\left\{\sum_{t=1}^{T} \frac{a\boldsymbol{\phi}^{\mathrm{T}}}{\boldsymbol{\phi}^{\mathrm{T}}\hat{\boldsymbol{\Sigma}}\boldsymbol{\phi}} Q^{\pi}\right\} \equiv E\left\{\sum_{t=1}^{T} \frac{\left((\boldsymbol{\theta} + \boldsymbol{\varepsilon}_t)^{\mathrm{T}}\boldsymbol{\phi}\right)\boldsymbol{\phi}^{\mathrm{T}}}{\boldsymbol{\phi}^{\mathrm{T}}\hat{\boldsymbol{\Sigma}}\boldsymbol{\phi}} Q^{\pi}\right\} = E\left\{\sum_{t=1}^{T} \frac{(\boldsymbol{\theta}'^{\mathrm{T}}\boldsymbol{\phi})\boldsymbol{\phi}^{\mathrm{T}}}{\boldsymbol{\phi}^{\mathrm{T}}\hat{\boldsymbol{\Sigma}}\boldsymbol{\phi}} Q^{\pi}\right\}.$$

This equation yields

$$\boldsymbol{\theta}'^{\mathrm{T}} = E\left\{\sum_{t=1}^{T} \frac{\left((\boldsymbol{\theta} + \boldsymbol{\varepsilon}_t)^{\mathrm{T}}\boldsymbol{\phi}\right)\boldsymbol{\phi}^{\mathrm{T}}}{\boldsymbol{\phi}^{\mathrm{T}}\hat{\boldsymbol{\Sigma}}\boldsymbol{\phi}} Q^{\pi}\right\} E\left\{\sum_{t=1}^{T} \frac{\boldsymbol{\phi}\boldsymbol{\phi}^{\mathrm{T}}}{\boldsymbol{\phi}^{\mathrm{T}}\hat{\boldsymbol{\Sigma}}\boldsymbol{\phi}} Q^{\pi}\right\}^{-1}$$

$$= \boldsymbol{\theta}^{\mathrm{T}} + E\left\{\sum_{t=1}^{T} \frac{\boldsymbol{\varepsilon}_t^{\mathrm{T}}\boldsymbol{\phi}\boldsymbol{\phi}^{\mathrm{T}}}{\boldsymbol{\phi}^{\mathrm{T}}\hat{\boldsymbol{\Sigma}}\boldsymbol{\phi}} Q^{\pi}\right\} E\left\{\sum_{t=1}^{T} \frac{\boldsymbol{\phi}\boldsymbol{\phi}^{\mathrm{T}}}{\boldsymbol{\phi}^{\mathrm{T}}\hat{\boldsymbol{\Sigma}}\boldsymbol{\phi}} Q^{\pi}\right\}^{-1}$$

and finally with $\mathbf{W} = \boldsymbol{\phi}\boldsymbol{\phi}^{\mathrm{T}}(\boldsymbol{\phi}^{\mathrm{T}}\hat{\boldsymbol{\Sigma}}\boldsymbol{\phi})^{-1}$ we get

$$\boldsymbol{\theta}' = \boldsymbol{\theta} + E\left\{\sum_{t=1}^{T}\mathbf{W}Q^{\pi}\right\}^{-1} E\left\{\sum_{t=1}^{T}\mathbf{W}\boldsymbol{\varepsilon}_t Q^{\pi}\right\}.$$

If $\hat{\boldsymbol{\Sigma}}$ is diagonal, that is, the exploration of the parameters is pairwise independent, all parameters employ the same exploration, and the exploration is constant over rollouts, $\mathbf{W}$ simplifies to $\mathbf{W} = \boldsymbol{\phi}\boldsymbol{\phi}^{\mathrm{T}}(\boldsymbol{\phi}^{\mathrm{T}}\boldsymbol{\phi})^{-1}$. Normalized basis functions $\boldsymbol{\phi}$ further simplify $\mathbf{W}$ to $\mathbf{W}\left(\mathbf{s}, t\right) = \boldsymbol{\phi}\boldsymbol{\phi}^{\mathrm{T}}$.

If only one parameter is active at each time step, $\mathbf{W}\left(\mathbf{s}, t\right)$ is diagonal and Equation (4.A.3) simplifies to

$$\theta_i' = \theta_i + \frac{E\left\{\sum_{t=1}^{T}\phi_i^2/\left(\boldsymbol{\phi}^{\mathrm{T}}\hat{\boldsymbol{\Sigma}}\boldsymbol{\phi}\right)\varepsilon_{i,t}Q^{\pi}\right\}}{E\left\{\sum_{t=1}^{T}\phi_i^2/\left(\boldsymbol{\phi}^{\mathrm{T}}\hat{\boldsymbol{\Sigma}}\boldsymbol{\phi}\right)Q^{\pi}\right\}} \tag{4.2}$$

$$= \theta_i + \frac{E\left\{\sum_{t=1}^{T}\sigma_i^{-1}\varepsilon_{i,t}Q^{\pi}\right\}}{E\left\{\sum_{t=1}^{T}\sigma_i^{-1}Q^{\pi}\right\}}, \tag{4.3}$$

where θ_i' is one individual parameter, ϕ_i and $\varepsilon_{i,t}$ are the corresponding elements of $\boldsymbol{\phi}$ and $\boldsymbol{\varepsilon}_t$, and σ_i is the respective entry of the diagonal of $\hat{\boldsymbol{\Sigma}}$. If the σ_i are constant over the rollouts we get $\theta_i' = \theta_i + E\{\sum_{t=1}^{T}\varepsilon_{i,t}Q^{\pi}\}/E\{\sum_{t=1}^{T}Q^{\pi}\}$. The independence simplification in Equations (4.2, 4.3) works well in practice, even if there is some overlap between the activations, such as in the case of dynamical system motor primitives (Ijspeert et al, 2002b,a; Schaal et al, 2003, 2007). Weighting the exploration with the basis functions, as in Equation (4.2), yields slightly better results than completely ignoring the interactions, as in Equation (4.3).

The policy can be equivalently expressed as

$$\pi\left(a_t|\mathbf{s}_t,t\right) = p\left(a_t|\mathbf{s}_t,t,\boldsymbol{\varepsilon}_t\right)p\left(\boldsymbol{\varepsilon}_t|\mathbf{s}_t,t\right) = p\left(a_t|\mathbf{s}_t,t,\boldsymbol{\varepsilon}_t\right)\mathcal{N}\left(\boldsymbol{\varepsilon}_t|\mathbf{0},\hat{\boldsymbol{\Sigma}}\right).$$

Applying Equation (4.2.2) to the variance $\hat{\boldsymbol{\Sigma}}$ we get

$$\partial_{\hat{\boldsymbol{\Sigma}}'}L_{\hat{\boldsymbol{\Sigma}}}\left(\hat{\boldsymbol{\Sigma}}'\right) = E\left\{\sum_{t=1}^{T}\partial_{\hat{\boldsymbol{\Sigma}}'}\log\mathcal{N}\left(\boldsymbol{\varepsilon}_t|\mathbf{0},\hat{\boldsymbol{\Sigma}}'\right)Q^{\pi}\right\},$$

as $p\left(a_t|\mathbf{s}_t,t,\boldsymbol{\varepsilon}_t\right)$ is independent from the variance. Setting this equation to zero and solving for $\hat{\boldsymbol{\Sigma}}'$ yields

$$\hat{\boldsymbol{\Sigma}}' = \frac{E\left\{\sum_{t=1}^{T}\boldsymbol{\varepsilon}_t\boldsymbol{\varepsilon}_t^{\mathsf{T}}Q^{\pi}\right\}}{E\left\{\sum_{t=1}^{T}Q^{\pi}\right\}},$$

which is the same solution as we get in standard maximum likelihood problems.

The same derivation holds if we use Equation (4.2.2) instead of Equation (4.2.2). Then, the state-action value function Q^{π} is replaced everywhere by the return R.

5

Reinforcement Learning to Adjust Parametrized Motor Primitives to New Situations

with Andreas Wilhelm[*] and Erhan Oztop[†]

Summary. Humans manage to adapt learned movements very quickly to new situations by generalizing learned behaviors from similar situations. In contrast, robots currently often need to re-learn the complete movement. In this chapter, we propose a method that learns to generalize parametrized motor plans by adapting a small set of global parameters, called meta-parameters. We employ reinforcement learning to learn the required meta-parameters to deal with the current situation, described by states. We introduce an appropriate reinforcement learning algorithm based on a kernelized version of the reward-weighted regression. To show its feasibility, we evaluate this algorithm on a toy example and compare it to several previous approaches. Subsequently, we apply the approach to three robot tasks, i.e., the generalization of throwing movements in darts, of hitting movements in table tennis, and of throwing balls where the tasks are learned on several different real physical robots, i.e., a Barrett WAM, a BioRob, the JST-ICORP/SARCOS CBi and a Kuka KR 6.

5.1 Introduction

Human movements appear to be represented using movement templates, also called motor primitives (Schmidt and Wrisberg, 2000). Once learned, these templates allow humans to quickly adapt their movements to variations of the situation without the need of re-learning the complete movement. For example, the overall shape of table tennis forehands are very similar when

[*] University of Applied Sciences Ravensburg-Weingarten, Doggenriedstr., 88250 Weingarten, Germany.

[†] Özyeğin University Robotics Laboratory, Çekmeköy Campus, Nişantepe District, Orman Street, 34794 Çekmeköy - İstanbul, Turkey.

J. Kober and J. Peters, *Learning Motor Skills*,
Springer Tracts in Advanced Robotics 97,
DOI: 10.1007/978-3-319-03194-1_5, © Springer International Publishing Switzerland 2014

the swing is adapted to varied trajectories of the incoming ball and a different targets on the opponent's court. To accomplish such behavior, the human player has learned by trial and error how the global parameters of a generic forehand need to be adapted due to changes in the situation (Muelling et al, 2010, 2011).

In robot learning, motor primitives based on dynamical systems (Ijspeert et al, 2002a; Schaal et al, 2007) can be considered a technical counterpart to these templates. They allow acquiring new behaviors quickly and reliably both by imitation and reinforcement learning. Resulting successes have shown that it is possible to rapidly learn motor primitives for complex behaviors such as tennis-like swings (Ijspeert et al, 2002a), T-ball batting (Peters and Schaal, 2008c), drumming (Pongas et al, 2005), biped locomotion (Nakanishi et al, 2004), ball-in-a-cup (Kober and Peters, 2011b), and even in tasks with potential industrial applications (Urbanek et al, 2004). While the examples are impressive, they do not yet address how a motor primitive can be generalized to a different behavior by trial and error without re-learning the task. Such generalization of behaviors can be achieved by adapting the meta-parameters of the movement representation. Meta-parameters are defined as a small set of parameters that adapt the global movement behavior. The dynamical system motor primitives can be adapted both spatially and temporally without changing the overall shape of the motion (Ijspeert et al, 2002a). In this chapter, we learn a mapping from a range of changed situations, described by states, to the meta-parameters to adapt the template's behavior. We consider movements where it is sufficient to reshape (e.g., rescale the motion spatially and/or temporally) the global movement by optimizing meta-parameters to adapt to a new situation instead of tuning the movement primitive's shape parameters that describe the fine details of the movement.

Dynamical systems motor primitives have the capability to adapt the movement to a changed end positions. Here, the end position is a meta-parameter. This was exploited in (Ijspeert et al, 2002a) for tennis-like swings with static ball targets and in (Pastor et al, 2009) for object manipulation. In these papers, the desired end position is given in Cartesian coordinates and the movement primitives operate in Cartesian coordinates as well. Thus, the meta-parameters of the motor primitives are straightforward to set. In this chapter, we are interested in non-intuitive connections, where the relation between the desired outcome and the meta-parameters is not straightforward. There is related prior work in the context of programming by demonstration by Ude et al (2010) and Kronander et al (2011) who employ supervised learning to learn a mapping from desired outcomes to meta-parameters for tasks such as reaching, throwing, drumming, and mini-golf. They assume that a teacher has presented a number of demonstrations that cannot be contradictory and the task is to imitate and generalize these demonstrations. Lampariello et al (2011) employ a global planner to provide demonstrations of optimal catching meta-parameters and use supervised learning approaches to generalize these in real-time. In contrast, in our setting the robot actively explores different movements and improves the behavior according to a cost function. It can deal with contradictory demonstrations and actively generate its own scenarios by exploration combined with self-improvement. As mentioned in (Ude et al, 2010), the two approaches may even be complimentary: reinforcement learning can provide demonstrations for supervised learning, and supervised learning can be used as a starting point for reinforcement learning.

Adapting movements to situations is also discussed in (Jetchev and Toussaint, 2009) in a supervised learning setting. Their approach is based on predicting a trajectory from a previously demonstrated set and refining it by motion planning. The authors note that kernel ridge regression performed poorly for the prediction if the new situation is far from previously seen ones as the algorithm yields the global mean. In our approach, we employ a cost weighted mean that overcomes this problem. If the situation is far from previously seen ones, large exploration will help to find a solution.

In machine learning, there have been many attempts to use meta-parameters in order to generalize between tasks (Caruana, 1997). Particularly, in grid-world domains, significant speed-up could be achieved by adjusting policies by modifying their meta-parameters, e.g., re-using options with different subgoals (McGovern and Barto, 2001). The learning of meta-parameters of the learning algorithm has been proposed as a model for neuromodulation in the brain (Doya, 2002). In contrast, we learn the meta-parameters of a motor skill in this chapter. In robotics, such meta-parameter learning could be particularly helpful due to the complexity of reinforcement learning for complex motor skills with high dimensional states and actions. The cost of experience is high as sample generation is time consuming and often requires human interaction (e.g., in cart-pole, for placing the pole back on the robot's hand) or supervision (e.g., for safety during the execution of the trial). Generalizing a teacher's demonstration or a previously learned policy to new situations may reduce both the complexity of the task and the number of required samples. Hence, a reinforcement learning method for acquiring and refining meta-parameters of pre-structured primitive movements becomes an essential next step, which we will address in this chapter.

This chapter does not address the problem of deciding whether it is more advantageous to generalize existing generic movements or to learn a novel one. Similar to most reinforcement learning approaches, the states and meta-parameters (which correspond to actions in the standard reinforcement learning settings) as well as the cost or reward function need to be designed by the user prior to the learning process. Here, we can only provide a few general indications with regard to the choice of these setting. Cost functions need to capture the desired outcome of the reinforcement learning process. Often the global target can be described verbally - but it is not obvious how the cost needs to be scaled and how to take secondary optimization criteria into account. For example, when throwing at a target, the global goal is hitting it. However, it is not always obvious which distance metric should be used to score misses, which secondary criteria (e.g. required torques) should be included, and which weight each criterion should be assigned. These choices influence both the learning performance and the final policy. Even for human reaching movements, the underlying cost function is not completely understood (Bays and Wolpert, 2007). In practice, informative cost functions (i.e., cost functions that contain a notion of closeness) often perform better than binary reward functions in robotic tasks. In this chapter, we used a number of cost functions both with and without secondary objectives. In the future, inverse reinforcement learning (Russell, 1998) may be a useful alternative to automatically recover underlying cost functions from data as done already in other settings.

The state of the environment needs to enable the robot to obtain sufficient information to react appropriately. The proposed algorithm can cope with superfluous states at a cost of slower learning. Similarly, the meta-parameters are defined by the underlying representation of the movement. For example, the dynamical systems motor primitives (Ijspeert et al, 2002a; Schaal et al, 2007) have meta-parameters for scaling the duration and amplitude of the movement as well as the possibility to change the final position. Restricting the meta-parameters to task relevant ones, may often speed up the learning process.

We present current work on automatic meta-parameter acquisition for motor primitives by reinforcement learning. We focus on learning the mapping from situations to meta-parameters and how to employ these in dynamical systems motor primitives. We extend the motor primitives of Ijspeert et al (2002a) with a learned meta-parameter function and re-frame the problem as an episodic reinforcement learning scenario. In order to obtain an algorithm for fast reinforcement learning of meta-parameters, we view reinforcement learning as a reward-weighted self-imitation (Peters and Schaal, 2008a; Kober and Peters, 2011b).

To have a general meta-parameter learning, we adopted a parametric method, the reward-weighed regression (Peters and Schaal, 2008a), and turned it into a non-parametric one. We call this method Cost-regularized Kernel Regression (CrKR), which is related to Gaussian process regression (Rasmussen and Williams, 2006) but differs in the key aspects of incorporating costs and exploration naturally. We compare the CrKR with a traditional policy gradient algorithm (Peters and Schaal, 2008c), the reward-weighted regression (Peters and Schaal, 2008a), and supervised learning (Ude et al, 2010; Kronander et al, 2011) on a toy problem in order to show that it outperforms available previously developed approaches. As complex motor control scenarios, we evaluate the algorithm in the acquisition of flexible motor primitives for dart games such as *Around the Clock* (Masters Games Ltd., 2010), for *table tennis*, and for *ball target throwing*.

5.2   Meta-Parameter Learning for Motor Primitives

The goal of this chapter is to show that elementary movements can be generalized by modifying only the meta-parameters of the primitives using learned mappings based on self-improvement. In Section 5.2.1, we first review how a single primitive movement can be represented and learned. We discuss how meta-parameters may be able to adapt the motor primitive spatially and temporally to the new situation. In order to develop algorithms that learn to automatically adjust such motor primitives, we model meta-parameter self-improvement as an episodic reinforcement learning problem in Section 5.2.2. While this problem could in theory be treated with arbitrary reinforcement learning methods, the availability of few samples suggests that more efficient, task appropriate reinforcement learning approaches are needed. To avoid the limitations of parametric function approximation, we aim for a kernel-based approach. When a movement is generalized, new parameter settings need to be explored. Hence, a predictive distribution over the meta-parameters is required to serve as an exploratory policy. These requirements lead to the method which we derive in Section 5.2.3 and employ for meta-parameter learning in Section 5.2.4.

5.2.1   Motor Primitives with Meta-Parameters

In this section, we review how the dynamical systems motor primitives (Ijspeert et al, 2002a; Schaal et al, 2007) discussed in Chapter 3 can be used for meta-parameter learning. The dynamical system motor primitives are a powerful movement representation that allows ensuring the stability of the movement[1], choosing between a rhythmic and a discrete movement and is invariant under rescaling of both duration and movement amplitude. These modification parameters can become part of the meta-parameters of the movement.

In this chapter, we focus on single stroke movements which appear frequently in human motor control (Wulf, 2007; Schaal et al, 2007). Therefore, we will always focus on the discrete version of the dynamical system motor primitives in this chapter.

The motor primitive policy is invariant under transformations of the initial position $\mathbf{x}_1^0$, the initial velocity $\mathbf{x}_2^0$, the goal $\mathbf{g}$, the goal velocity $\dot{\mathbf{g}}$, the amplitude $\mathbf{A}$, and the duration T as discussed in Chapter 3. These six modification parameters can be used as the meta-parameters $\boldsymbol{\gamma}$ of the movement. Obviously, we can make more use of the motor primitive framework by adjusting the meta-parameters $\boldsymbol{\gamma}$ depending on the current situation or state $\mathbf{s}$ according to a meta-parameter function $\bar{\boldsymbol{\gamma}}(\mathbf{s})$. The meta-parameter $\boldsymbol{\gamma}$ is

---

[1] Note that the dynamical systems motor primitives ensure the stability of the movement generation but cannot guarantee the stability of the movement execution (Ijspeert et al, 2002a; Schaal et al, 2007).

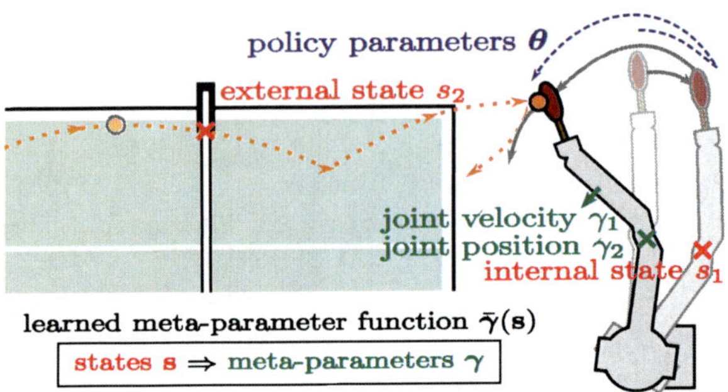

Fig. 5.1 This figure illustrates a table tennis task. The situation, described by the state **s**, corresponds to the positions and velocities of the ball and the robot at the time the ball is above the net. The meta-parameters γ are the joint positions and velocity at which the ball is hit. The policy parameters represent the backward motion and the movement on the arc. The meta-parameter function $\bar{\gamma}(\mathbf{s})$, which maps the state to the meta-parameters, is learned.

treated as a random variable where the variance correspond to the uncertainty. The state **s** can for example contain the current position, velocity and acceleration of the robot and external objects, as well as the target to be achieved. This paper focuses on learning the meta-parameter function $\bar{\gamma}(\mathbf{s})$ by episodic reinforcement learning.

Illustrations of the Learning Problem

We discuss the resulting learning problem based on the two examples shown in Figures 5.1 and 5.2.

As a first illustration of the meta-parameter learning problem, we take a table tennis task which is illustrated in Figure 5.1 (in Section 5.3.3, we will expand this example to a robot application). Here, the desired skill is to return a table tennis ball. The motor primitive corresponds to the hitting movement. When modeling a single hitting movement with dynamical-systems motor primitives (Ijspeert et al, 2002a), the combination of retracting and hitting motions would be represented by one movement primitive and can be learned by determining the movement parameters $\boldsymbol{\theta}$. These parameters can either be estimated by imitation learning or acquired by reinforcement learning. The return can be adapted by changing the paddle position and velocity at the hitting point. These variables can be influenced by modifying the meta-parameters of the motor primitive such as the final joint positions and velocities. The state consists of the current positions and velocities of the ball and the robot at the time the ball is directly above the net. The meta-parameter function $\bar{\gamma}(\mathbf{s})$ maps the state (the state of the ball and the robot before the return) to the meta-parameters γ (the final positions and velocities

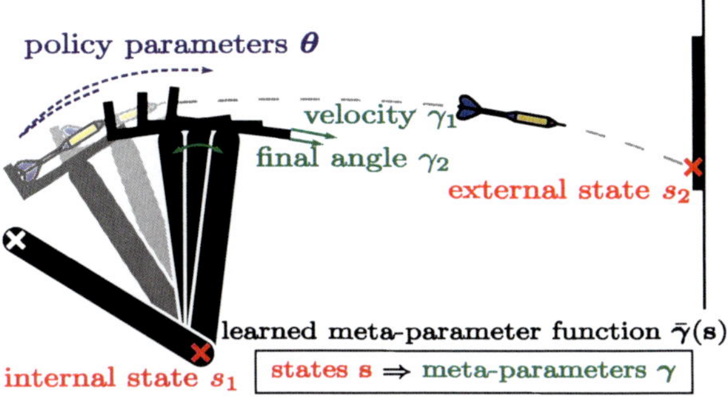

Fig. 5.2 This figure illustrates a 2D dart throwing task. The situation, described by the state **s** corresponds to the relative height. The meta-parameters γ are the velocity and the angle at which the dart leaves the launcher. The policy parameters represent the backward motion and the movement on the arc. The meta-parameter function $\bar{\gamma}(\mathbf{s})$, which maps the state to the meta-parameters, is learned.

of the motor primitive). Its variance corresponds to the uncertainty of the mapping.

In a 2D dart throwing task with a dart on a launcher which is illustrated in Figure 5.2 (in Section 5.3.2, we will expand this example to a robot application) the desired skill is to hit a specified point on a wall with a dart. The dart is placed on the launcher and held there by friction. The motor primitive corresponds to the throwing of the dart. When modeling a single dart's movement with dynamical-systems motor primitives (Ijspeert et al, 2002a), the combination of retracting and throwing motions would be represented by the movement parameters $\boldsymbol{\theta}$ of one movement primitive. The dart's impact position can be adapted to a desired target by changing the velocity and the angle at which the dart leaves the launcher. These variables can be influenced by changing the meta-parameters of the motor primitive such as the final position of the launcher and the duration of the throw. The state consists of the current position of the hand and the desired position on the target. If the thrower is always at the same distance from the wall the two positions can be equivalently expressed as the vertical distance. The meta-parameter function $\bar{\gamma}(\mathbf{s})$ maps the state (the relative height) to the meta-parameters γ (the final position **g** and the duration of the motor primitive T).

The approach presented in this chapter is applicable to any movement representation that has meta-parameters, i.e., a small set of parameters that allows to modify the movement. In contrast to (Lampariello et al, 2011; Jetchev and Toussaint, 2009; Grimes and Rao, 2008; Bentivegna et al, 2004b) our approach does not require explicit (re-)planning of the motion.

In the next sections, we derive and apply an appropriate reinforcement learning algorithm.

5.2.2 *Problem Statement: Meta-Parameter Self-improvement*

The problem of meta-parameter learning is to find a stochastic policy $\pi(\gamma|\mathbf{x}) = p(\gamma|\mathbf{s})$ that maximizes the expected return

$$J(\pi) = \int_{\mathbb{S}} p(\mathbf{s}) \int_{\mathbb{G}} \pi(\gamma|\mathbf{s}) R(\mathbf{s}, \gamma) d\gamma \, d\mathbf{s},$$

where $\mathbb{S}$ denotes the the space of states $\mathbf{s}$, $\mathbb{G}$ denotes the the space of meta-parameters γ, and $R(\mathbf{s}, \gamma)$ denotes all the rewards following the selection of the meta-parameter γ according to a situation described by state $\mathbf{s}$. Such a policy $\pi(\gamma|\mathbf{x})$ is a probability distribution over meta-parameters given the current state. The stochastic formulation allows a natural incorporation of exploration, and the optimal time-invariant policy has been shown to be stochastic in the case of hidden state variables (Sutton et al, 1999; Jaakkola et al, 1993). The return of an episode is $R(\mathbf{s}, \gamma) = T^{-1} \sum_{t=0}^{T} r^t$ with number of steps T and rewards r^t. For a parametrized policy π with parameters $\mathbf{w}$ it is natural to first try a policy gradient approach such as finite-difference methods, vanilla policy gradient approaches and natural gradients. While we will denote the shape parameters by $\boldsymbol{\theta}$, we denote the parameters of the meta-parameter function by $\mathbf{w}$. Reinforcement learning of the meta-parameter function $\bar{\gamma}(\mathbf{s})$ is not straightforward as only few examples can be generated on the real system and trials are often quite expensive. The credit assignment problem is non-trivial as the whole movement is affected by every change in the meta-parameter function. Early attempts using policy gradient approaches resulted in tens of thousands of trials even for simple toy problems, which is not feasible on a real system.

Dayan and Hinton (1997) showed that an immediate reward can be maximized by instead minimizing the Kullback-Leibler divergence $D(\pi(\gamma|\mathbf{s}) R(\mathbf{s}, \gamma) \| \pi'(\gamma|\mathbf{s}))$ between the reward-weighted policy $\pi(\gamma|\mathbf{s})$ and the new policy $\pi'(\gamma|\mathbf{s})$. As we are in an episodic setting, this form of optimization solves the considered problem. Williams (1992) suggested to use Gaussian noise in this context; hence, we employ a policy of the form

$$\pi(\gamma|\mathbf{s}) = \mathcal{N}(\gamma|\bar{\gamma}(\mathbf{s}), \sigma^2(\mathbf{s})\mathbf{I}),$$

where we have the deterministic mean policy $\bar{\gamma}(\mathbf{s}) = \phi(\mathbf{s})^{\mathrm{T}} \mathbf{w}$ with basis functions $\phi(\mathbf{s})$ and parameters $\mathbf{w}$ as well as the variance $\sigma^2(\mathbf{s})$ that determines the exploration $\epsilon \sim \mathcal{N}(\mathbf{0}, \sigma^2(\mathbf{s})\mathbf{I})$ as e.g., in (Peters and Schaal, 2008c). The parameters $\mathbf{w}$ can then be adapted by reward-weighted regression in an immediate reward (Peters and Schaal, 2008a) or episodic reinforcement learning scenario (see Chapter 4). The reasoning behind this reward-weighted regression is that the reward can be treated as an improper probability distribution over indicator variables determining whether the action is optimal or not.

Algorithm 5.1 Meta-Parameter Learning

Preparation steps:

  Learn one or more motor primitives by imitation and/or reinforcement learning (yields shape parameters $\boldsymbol{\theta}$).

  Determine initial state $\mathbf{s}^0$, meta-parameters $\boldsymbol{\gamma}^0$, and cost C^0 corresponding to the initial motor primitive.

  Initialize the corresponding matrices $\mathbf{S}, \boldsymbol{\Gamma}, \mathbf{C}$.

  Choose a kernel $\mathbf{k}, \mathbf{K}$.

  Set a scaling parameter λ.

for all iterations j **do**

  Determine the state $\mathbf{s}^j$ specifying the situation.

  Calculate the meta-parameters $\boldsymbol{\gamma}^j$ by:

    Determine the mean of each meta-parameter i $\bar{\gamma}_i(\mathbf{s}^j) = \mathbf{k}(\mathbf{s}^j)^{\mathrm{T}} (\mathbf{K} + \lambda \mathbf{C})^{-1} \boldsymbol{\Gamma}_i$,

    Determine the variance $\sigma^2(\mathbf{s}^j) = k(\mathbf{s}^j, \mathbf{s}^j) - \mathbf{k}(\mathbf{s}^j)^{\mathrm{T}} (\mathbf{K} + \lambda \mathbf{C})^{-1} \mathbf{k}(\mathbf{s}^j)$,

    Draw the meta-parameters from a Gaussian distribution $\boldsymbol{\gamma}^j \sim \mathcal{N}(\boldsymbol{\gamma} | \bar{\boldsymbol{\gamma}}(\mathbf{s}^j), \sigma^2(\mathbf{s}^j)\mathbf{I})$

  Execute the motor primitive using the new meta-parameters.

  Calculate the cost c^j at the end of the episode.

  Update $\mathbf{S}, \boldsymbol{\Gamma}, \mathbf{C}$ according to the achieved result.

end for

5.2.3   *A Task-Appropriate Reinforcement Learning Algorithm*

Designing good basis functions is challenging, a nonparametric representation is better suited in this context. There is an intuitive way of turning the reward-weighted regression into a Cost-regularized Kernel Regression. The kernelization of the reward-weighted regression can be done straightforwardly (similar to Section 6.1 of (Bishop, 2006) for regular supervised learning). Inserting the reward-weighted regression solution $\mathbf{w} = (\boldsymbol{\Phi}^{\mathrm{T}} \mathbf{R} \boldsymbol{\Phi} + \lambda \mathbf{I})^{-1} \boldsymbol{\Phi}^{\mathrm{T}} \mathbf{R} \boldsymbol{\Gamma}_i$ and using the Woodbury formula[2] (Welling, 2010), we transform reward-weighted regression into a Cost-regularized Kernel Regression

---

[2] The equality $(\boldsymbol{\Phi}^{\mathrm{T}} \mathbf{R} \boldsymbol{\Phi} + \lambda \mathbf{I})^{-1} \boldsymbol{\Phi}^{\mathrm{T}} \mathbf{R} = \boldsymbol{\Phi}^{\mathrm{T}} (\boldsymbol{\Phi} \boldsymbol{\Phi}^{\mathrm{T}} + \lambda \mathbf{R}^{-1})^{-1}$ is straightforward to verify by left and right multiplying the non-inverted terms: $\boldsymbol{\Phi}^{\mathrm{T}} \mathbf{R}(\boldsymbol{\Phi} \boldsymbol{\Phi}^{\mathrm{T}} + \lambda \mathbf{R}^{-1}) = (\boldsymbol{\Phi}^{\mathrm{T}} \mathbf{R} \boldsymbol{\Phi} + \lambda \mathbf{I})\boldsymbol{\Phi}^{\mathrm{T}}$.

$$\bar{\gamma}_i = \phi(\mathbf{s})^{\mathrm{T}}\mathbf{w} = \phi(\mathbf{s})^{\mathrm{T}}\left(\mathbf{\Phi}^{\mathrm{T}}\mathbf{R}\mathbf{\Phi} + \lambda\mathbf{I}\right)^{-1}\mathbf{\Phi}^{\mathrm{T}}\mathbf{R}\mathbf{\Gamma}_i$$

$$= \phi(\mathbf{s})^{\mathrm{T}}\mathbf{\Phi}^{\mathrm{T}}\left(\mathbf{\Phi}\mathbf{\Phi}^{\mathrm{T}} + \lambda\mathbf{R}^{-1}\right)^{-1}\mathbf{\Gamma}_i, \tag{5.1}$$

where the rows of $\mathbf{\Phi}$ correspond to the basis functions $\phi(\mathbf{s}_i) = \mathbf{\Phi}_i$ of the training examples, $\mathbf{\Gamma}_i$ is a vector containing the training examples for meta-parameter component γ_i, and λ is a ridge factor. Next, we assume that the accumulated rewards R_k are strictly positive $R_k > 0$ and can be transformed into costs by $c_k = 1/R_k$. Hence, we have a cost matrix $\mathbf{C} = \mathbf{R}^{-1} = \mathrm{diag}(R_1^{-1}, \ldots, R_n^{-1})$ with the cost of all n data points. After replacing $\mathbf{k}(\mathbf{s}) = \phi(\mathbf{s})^{\mathrm{T}}\mathbf{\Phi}^{\mathrm{T}}$ and $\mathbf{K} = \mathbf{\Phi}\mathbf{\Phi}^{\mathrm{T}}$, we obtain the Cost-regularized Kernel Regression

$$\bar{\gamma}_i = \bar{\gamma}_i(\mathbf{s}) = \mathbf{k}(\mathbf{s})^{\mathrm{T}}\left(\mathbf{K} + \lambda\mathbf{C}\right)^{-1}\mathbf{\Gamma}_i,$$

which gives us a deterministic policy. Here, costs correspond to the uncertainty about the training examples. Thus, a high cost is incurred for being further away from the desired optimal solution at a point. In our formulation, a high cost therefore corresponds to a high uncertainty of the prediction at this point.

In order to incorporate exploration, we need to have a stochastic policy and, hence, we need a predictive distribution. This distribution can be obtained by performing the policy update with a Gaussian process regression and we directly see from the kernel ridge regression that

$$\sigma^2(\mathbf{s}) = k(\mathbf{s}, \mathbf{s}) + \lambda - \mathbf{k}(\mathbf{s})^{\mathrm{T}}\left(\mathbf{K} + \lambda\mathbf{C}\right)^{-1}\mathbf{k}(\mathbf{s}),$$

where $k(\mathbf{s}, \mathbf{s}) = \phi(\mathbf{s})^{\mathrm{T}}\phi(\mathbf{s})$ is the norm of the point in the kernel space. We call this algorithm Cost-regularized Kernel Regression. Algorithm 5.1 describes the complete learning procedure, where the rows of $\mathbf{S}$ correspond to the states of the training examples $\mathbf{s}_i = \mathbf{S}_i$.

The algorithm corresponds to a Gaussian process regression where the costs on the diagonal are input-dependent noise priors. The parameter λ acts as a exploration-exploitation trade-off parameter as illustrated in Figure 5.5. Gaussian processes have been used previously for reinforcement learning (Engel et al, 2005) in value function based approaches while here we use them to learn the policy.

5.2.4 *Meta-Parameter Learning by Reinforcement Learning*

As a result of Section 5.2.3, we have a framework of motor primitives as introduced in Section 5.2.1 that we can use for reinforcement learning of meta-parameters as outlined in Section 5.2.2. We have generalized the reward-weighted regression policy update to instead become a Cost-regularized Kernel Regression (CrKR) update where the predictive variance is used for exploration. In Algorithm 1, we show the complete algorithm resulting from these steps.

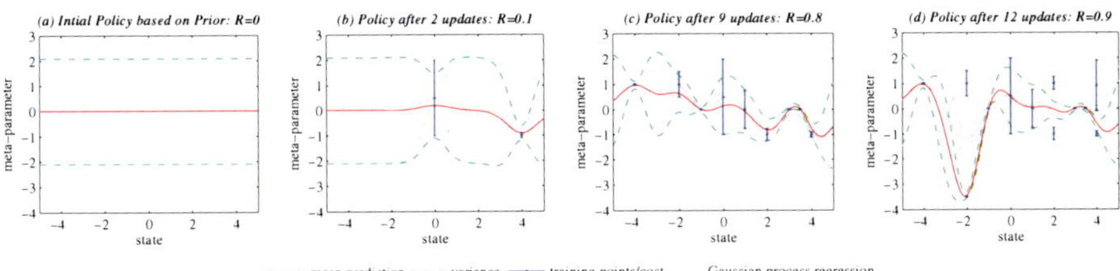

Fig. 5.3 This figure illustrates the meaning of policy improvements with Cost-regularized Kernel Regression. Each sample consists of a state, a meta-parameter and a cost where the cost is indicated the blue error bars. The red line represents the improved mean policy, the dashed green lines indicate the exploration/variance of the new policy. For comparison, the gray lines show standard Gaussian process regression. As the cost of a data point is equivalent to having more noise, pairs of states and meta-parameter with low cost are more likely to be reproduced than others with high costs.

The algorithm receives three inputs, i.e., (i) a motor primitive that has associated meta-parameters γ, (ii) an initial example containing state s^0, meta-parameter γ^0 and cost C^0, as well as (iii) a scaling parameter λ. The initial motor primitive can be obtained by imitation learning (Ijspeert et al, 2002a) and, subsequently, improved by parametrized reinforcement learning algorithms such as policy gradients (Peters and Schaal, 2008c) or Policy learning by Weighting Exploration with the Returns (PoWER) [Kober and Peters, 2011b, Chapter 4]. The demonstration also yields the initial example needed for meta-parameter learning. While the scaling parameter is an open parameter, it is reasonable to choose it as a fraction of the average cost and the output noise parameter (note that output noise and other possible hyper-parameters of the kernel can also be obtained by approximating the unweighted meta-parameter function).

Illustration of the Algorithm

In order to illustrate this algorithm, we will use the example of the table tennis task introduced in Section 5.2.1. Here, the robot should hit the ball accurately while not destroying its mechanics. Hence, the cost could correspond to the distance between the ball and the paddle, as well as the squared torques. The initial policy is based on a prior, illustrated in Figure 5.3(a), that has a variance for initial exploration (it often makes sense to start with a uniform prior). This variance is used to enforce exploration. To return a ball, we sample the meta-parameters from the policy based on the current state. After the trial the cost is determined and, in conjunction with the employed meta-parameters, used to update the policy. If the cost is large (e.g., the ball was far from the racket), the variance of the policy is large as it may still be improved and therefore needs exploration. Furthermore, the mean of the policy is shifted only slightly towards the observed example as we are uncertain about the optimality of this action. If the cost is small, we know that we are close to an optimal policy (e.g., the racket hit the ball off-center) and only have to search in a small region around the observed trial. The effects of the cost on the mean and the variance are illustrated in Figure 5.3(b). Each additional sample refines the policy and the overall performance improves (see Figure 5.3(c)). If a state is visited several times and different meta-parameters are sampled, the policy update must favor the meta-parameters with lower costs. If several sets of meta-parameters have similarly low costs, where it converges depends on the order of samples. The cost function should be designed to avoid this behavior and to favor a single set. The exploration has to be restricted to safe meta-parameter ranges. Algorithm 1 exhibits this behavior as the exploration is only local and restricted by the prior (see Figure 5.3). If the initial policy is safe, exploring the neighboring regions is likely to be safe as well. Additionally, lower level controllers as well as the mechanics of the robot ensure that kinematic and dynamic constrains are satisfied and a term in the cost function can be used to discourage potentially harmful movements.

In the example of the 2D dart throwing task, the cost is similar. Here, the robot should throw darts accurately while not destroying its mechanics. Hence, the cost could correspond to the error between desired goal and the impact point, as well as the absolute velocity of the end-effector. Often the state is determined by the environment, e.g., the ball trajectory in table tennis depends on the opponent. However, for the dart setting, we could choose the next target and thus employ CrKR as an active learning approach by picking states with large variances. In the dart throwing example we have a correspondence between the state and the outcome similar to a regression problem. However, the mapping between the state and the meta-parameter is not unique. The same height can be achieved by different combinations of velocities and angles. Averaging these combinations is likely to generate inconsistent solutions. The regression must hence favor the meta-parameters with the lower costs. CrKR can be employed as a regularized regression method in this setting.

5.3   Evaluations and Experiments

In Section 5.2, we have introduced both a framework for meta-parameter self-improvement as well as an appropriate reinforcement learning algorithm used in this framework. In this section, we will first show that the presented reinforcement learning algorithm yields higher performance than off-the shelf approaches. Hence, we compare it on a simple planar cannon shooting problem (Lawrence et al, 2003) with the preceding reward-weighted regression, an off-the-shelf finite difference policy gradient approach, and show the advantages over supervised learning approaches.

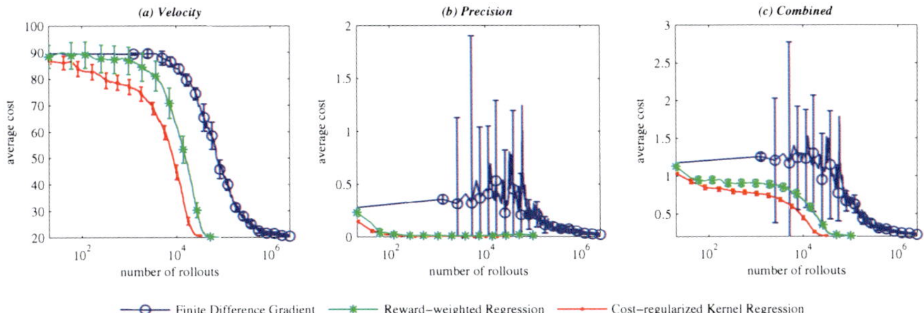

Fig. 5.4 This figure shows the performance of the compared algorithms averaged over 10 complete learning runs. Cost-regularized Kernel Regression finds solutions with the same final performance two orders of magnitude faster than the finite difference gradient (FD) approach and twice as fast as the reward-weighted regression. At the beginning FD often is highly unstable due to our attempts of keeping the overall learning speed as high as possible to make it a stronger competitor. The lines show the median and error bars indicate standard deviation. The initialization and the initial costs are identical for all approaches. However, the omission of the first twenty rollouts was necessary to cope with the logarithmic rollout axis. The number of rollouts includes the rollouts not used to update the policy.

The resulting meta-parameter learning framework can be used in a variety of settings in robotics. We consider three scenarios here, i.e., (i) dart throwing with a simulated Barrett WAM, a real Kuka KR 6, and the JST-ICORP/SARCOS humanoid robot CBi (Cheng et al, 2007), (ii) table tennis with a simulated robot arm and a real Barrett WAM, and (iii) throwing a ball at targets with a MATLAB simulation and a real BioRob (Lens et al, 2010).

5.3.1   Benchmark Comparison: Toy Cannon Shots

In the first task, we only consider a simple simulated planar cannon shooting where we benchmark our Reinforcement Learning by Cost-regularized Kernel Regression approach against a finite difference gradient estimator and the reward-weighted regression. Additionally we contrast our reinforcement learning approach to a supervised one. Here, we want to learn an optimal policy for a 2D toy cannon environment similar to (Lawrence et al, 2003). This benchmark example serves to illustrate out approach and to compare it to various previous approaches.

The setup is given as follows: A toy cannon is at a fixed location $[0.0, 0.1]\, m$. The trajectory of the cannon ball depends on the angle with respect to the ground and the speed at which it leaves the cannon. The flight of the canon ball is simulated as ballistic flight of a point mass with Stokes's drag as wind model. The cannon ball is supposed to hit the ground at a

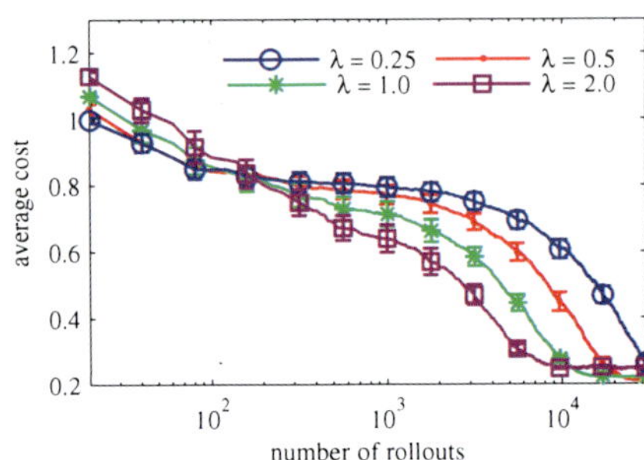

Fig. 5.5 This figure illustrates the influence of the parameter λ for the Cost-regularized Kernel Regression. The red curve ($\lambda = 0.5$) corresponds to the red curve (Cost-regularized Kernel Regression) in Figure 5.4(c). The parameter λ trades off the exploration versus the exploitation. A higher λ leads to larger exploration and, thus, faster convergence to a suboptimal solution. The results are averaged over 10 complete learning runs. The lines show the median and error bars indicate standard deviation. The number of rollouts includes the rollouts not used to update the policy.

desired distance. The desired distance $[1..3]\,m$ and the wind speed $[0..1]\,m/s$, which is always horizontal, are used as input states, the velocities in horizontal and vertical directions are the meta-parameters (which influences the angle and the speed of the ball leaving the cannon). In this benchmark we do not employ the motor primitives but set the meta-parameters directly. Lower speed can be compensated by a larger angle. Thus, there are different possible policies for hitting a target; we intend to learn the one which is optimal for a given cost function. This cost function is defined as

$$c = (b_x - s_x)^2 + 0.01 \left(\dot{b}_x^2 + \dot{b}_z^2 \right),$$

where b_x is the impact position on the ground, s_x the desired impact position as indicated by the state, and $\dot{b}_{\{x,z\}}$ are the horizontal and vertical velocities of the cannon ball at the impact point respectively. It corresponds to maximizing the precision while minimizing the employed energy according to the chosen weighting. The input states (desired distance and wind speed) are drawn from a uniform distribution and directly passed to the algorithms. All approaches performed well in this setting, first driving the position error to zero and, subsequently, optimizing the impact velocity. The experiment was initialized with $[1, 10]\,m/s$ as initial ball velocities and $1\,m/s$ as wind velocity. This setting corresponds to a very high parabola, which is far from optimal. For plots, we evaluate the policy on a test set of 25 uniformly randomly chosen points that remain the same throughout of the experiment and are never used in the learning process but only to generate Figure 5.4.

We compare our novel algorithm to a finite difference policy gradient (FD) method (Peters and Schaal, 2008c) and to the reward-weighted regression (RWR) (Peters and Schaal, 2008a). The FD method uses a parametric policy that employs radial basis functions in order to represent the policy and perturbs the parameters. We used 25 Gaussian basis functions on a regular grid for each meta-parameter, thus a total of 50 basis functions. The

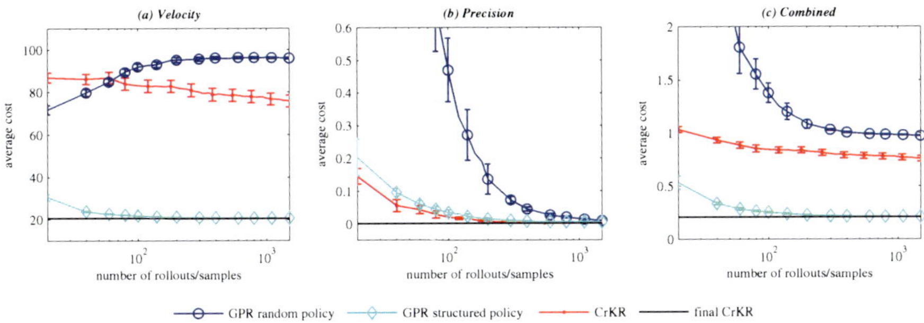

Fig. 5.6 In this figure, we compare Gaussian process regression (GPR) in a supervised learning setting as proposed by (Ude et al, 2010; Kronander et al, 2011) to Cost-regularized Kernel Regression (CrKR) in a reinforcement learning setting. The red curve corresponds to the red curve (Cost-regularized Kernel Regression) in Figure 5.4. The GPR is trained with samples from the prior used for the CrKR (blue line) and with samples of the final CrKR policy (cyan line) respectively. The black line indicates the cost after CrKR has converged. GPR with samples drawn from the final policy performs best. Please note that this comparison is contrived as the role of CrKR is to discover the policy that is provided to "GPR structured policy". GPR can only reproduce the demonstrated policy, which is achieved perfectly with 1000 samples. GPR can reproduce the demonstrated policy more accurately if more samples are available. However, it cannot improve the policy according to a cost function and it is impacted by contradictory demonstrations. The results are averaged over 10 complete learning runs. The lines show the median and error bars indicate standard deviation. The number of rollouts includes the rollouts not used to update the policy.

number of basis functions, the learning rate, as well as the magnitude of the perturbations were tuned for best performance. We used 51 sets of uniformly perturbed parameters for each update step. The perturbed policies were evaluated on a batch of 25 input parameters to avoid over-fitting on specific input states.The FD algorithm converges after approximately 2000 batch gradient evaluations, which corresponds to 2,550,000 shots with the toy cannon.

The RWR method uses the same parametric policy as the finite difference gradient method. Exploration is achieved by adding Gaussian noise to the mean policy. All open parameters were tuned for best performance. The reward transformation introduced by Peters and Schaal (2008a) did not improve performance in this episodic setting. The RWR algorithm converges after approximately 40,000 shots with the toy cannon. For the Cost-regularized Kernel Regression (CrKR) the inputs are chosen randomly from a uniform distribution. We use Gaussian kernels and the open parameters were optimized by cross-validation on a small test set prior to the experiment. Each trial is added as a new training point if it landed in the desired distance range. The CrKR algorithm converges after approximately 20,000 shots with the toy cannon. The bandwidth of the kernels used for CrKR is in the same

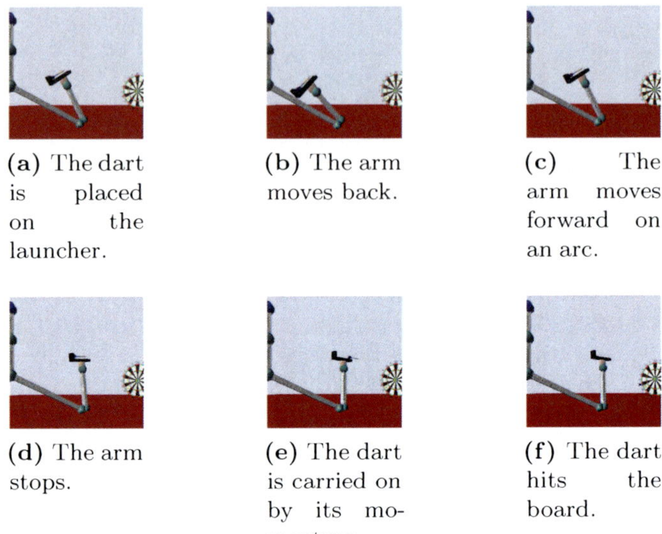

(a) The dart is placed on the launcher.

(b) The arm moves back.

(c) The arm moves forward on an arc.

(d) The arm stops.

(e) The dart is carried on by its momentum.

(f) The dart hits the board.

Fig. 5.7 This figure shows a dart throw in a physically realistic simulation

order of magnitude as the bandwidth of the basis functions. However, due to the non-parametric nature of CrKR, narrower kernels can be used to capture more details in order to improve performance. Figure 5.5 illustrates the influence of the parameter λ for the CrKR.

After convergence, the costs of CrKR are the same as for RWR and slightly lower than those of the FD method. The CrKR method needs two orders of magnitude fewer shots than the FD method. The RWR approach requires twice the shots of CrKR demonstrating that a non-parametric policy, as employed by CrKR, is better adapted to this class of problems than a parametric policy. The squared error between the actual and desired impact is approximately 5 times higher for the finite difference gradient method, see Figure 5.4.

Compared to standard Gaussian process regression (GPR) in a supervised setting, CrKR can improve the policy over time according to a cost function and outperforms GPR in settings where different combinations of meta-parameters yield the same result. For details, see Figure 5.6.

5.3.2 *Robot Dart-Throwing Games*

Now, we turn towards the complete framework, i.e., we intend to learn the meta-parameters for motor primitives in discrete movements. We compare the Cost-regularized Kernel Regression (CrKR) algorithm to the reward-weighted regression (RWR). As a sufficiently complex scenario, we chose a robot dart

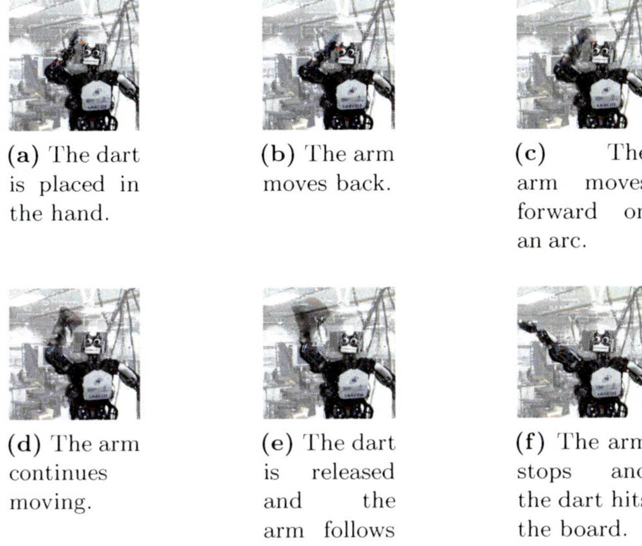

(a) The dart is placed in the hand.

(b) The arm moves back.

(c) The arm moves forward on an arc.

(d) The arm continues moving.

(e) The dart is released and the arm follows through.

(f) The arm stops and the dart hits the board.

Fig. 5.8 This figure shows a dart throw on the real JST-ICORP/SARCOS humanoid robot CBi

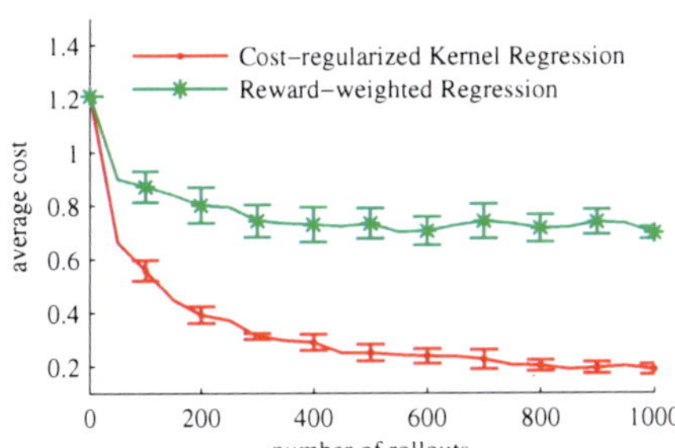

Fig. 5.9 This figure shows the cost function of the dart-throwing task for a whole game *Around the Clock* in each rollout. The costs are averaged over 10 runs with the error-bars indicating standard deviation. The number of rollouts includes the rollouts not used to update the policy.

throwing task inspired by (Lawrence et al, 2003). However, we take a more complicated scenario and choose dart games such as *Around the Clock* (Masters Games Ltd., 2010) instead of simple throwing at a fixed location. Hence, it will have an additional parameter in the state depending on the location on the dartboard that should come next in the sequence. The acquisition of a basic motor primitive is achieved using previous work on imitation learning (Ijspeert et al, 2002a). Only the meta-parameter function is learned using CrKR or RWR. For the learning process, the targets (which are part of the state) are uniformly distributed on the dartboard. For the evaluation the targets are placed in the center of the fields. The reward is calculated based on

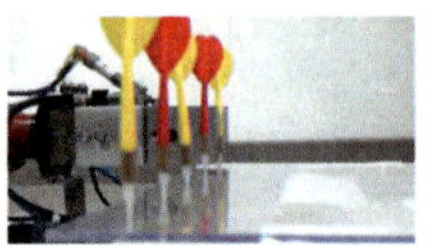   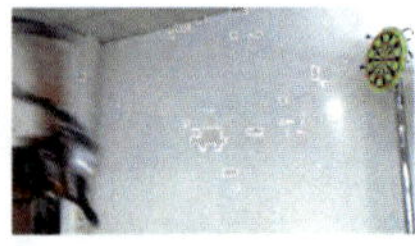

(a) The dart is picked up.

(b) The arm moves forward on an arc.

(c) The arm continues moving.

(d) The dart is released.

 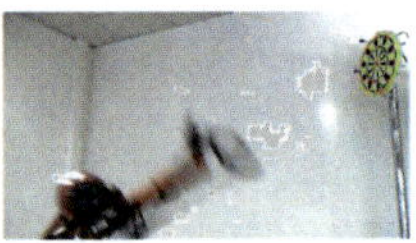  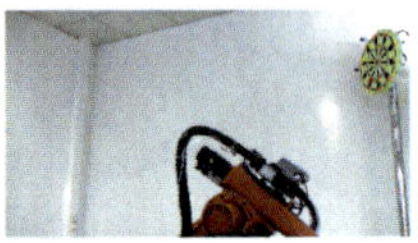

(e) The arm follows through.

(f) The arm continues moving.

(g) The arm returns to the pick-up position.

(h) The dart has hit the board.

Fig. 5.10 This figure shows a dart throw on the real Kuka KR 6 robot

the impact position observed by a vision system in the real robot experiments or the simulated impact position.

The dart is placed on a launcher attached to the end-effector and held there by stiction. We use the Barrett WAM robot arm in order to achieve the high accelerations needed to overcome the stiction. See Figure 5.7, for a complete throwing movement. The motor primitive is trained by imitation learning with kinesthetic teach-in. We use the Cartesian coordinates with respect to the center of the dart board as input states. In comparison with the benchmark example, we cannot directly influence the release velocity in this setup. Hence, we employ the parameter for the final position $\mathbf{g}$, the time scale of the motor primitive τ and the angle around the vertical axis (i.e., the orientation towards the dart board to which the robot moves before throwing) as meta-parameters instead. The popular dart game *Around the Clock* requires the player to hit the numbers in ascending order, then the bulls-eye. As energy is lost overcoming the stiction of the launching sled, the darts fly lower and we placed the dartboard lower than official rules require. The cost function is defined as

$$c = 10 \sqrt{\sum_{i \in \{x,z\}} (d_i - s_i)^2} + \tau,$$

where d_i are the horizontal and vertical positions of the dart on the dartboard after the throw, s_i are the horizontal and vertical positions of the target corresponding to the state, and τ corresponds to the velocity of the motion. After approximately 1000 throws the algorithms have converged but CrKR yields a high performance already much earlier (see Figure 5.9). We again used a parametric policy with radial basis functions for RWR. Here, we employed

225 Gaussian basis function on a regular grid per meta-parameter. Designing a good parametric policy proved very difficult in this setting as is reflected by the poor performance of RWR.

This experiment has also being carried out on three real, physical robots, i.e., a Barrett WAM, the humanoid robot CBi (JST-ICORP/SARCOS), and a Kuka KR 6. CBi was developed within the framework of the JST-ICORP Computational Brain Project at ATR Computational Neuroscience Labs. The hardware of the robot was developed by the American robotic development company SARCOS. CBi can open and close the fingers which helps for more human-like throwing instead of the launcher employed by the Barrett WAM. See Figure 5.8 for a throwing movement.

We evaluated the approach on a setup using the Kuka KR 6 robot and a pneumatic gripper. The robot automatically picks up the darts from a stand. The position of the first degree of freedom (horizontal position) as well as the position of the fifth degree of freedom and the release timing (vertical position) were controlled by the algorithm. Due to inaccurate release timing the vertical position varied in a range of $10cm$. Additionally the learning approach had to cope with non-stationary behavior as the outcome of the same set of parameters changed by one third of the dart board diameter upward. Despite these additional complications the robot learned to reliably (within the reproduction accuracy of $10cm$ as noted above) hit all positions on the dart board using only a total of 260 rollouts. See Figure 5.10 for a throwing movement.

5.3.3   Robot Table Tennis

In the second evaluation of the complete framework, we use the proposed method for hitting a table tennis ball in the air. The setup consists of a ball gun that serves to the forehand of the robot, a Barrett WAM and a standard sized table. The movement of the robot has three phases. The robot is in a rest posture and starts to swing back when the ball is launched. During this swing-back phase, the open parameters for the stroke are to be learned. The second phase is the hitting phase which ends with the contact of the ball and racket. In the final phase, the robot gradually ends the stroking motion and returns to the rest posture. See Figure 5.11 for an illustration of a complete episode and Chapter 3 for a more detailed description. The movements in the three phases are represented by three motor primitives obtained by imitation learning. We only learn the meta-parameters for the hitting phase.

The meta-parameters are the joint positions $\mathbf{g}$ and velocities $\dot{\mathbf{g}}$ for all seven degrees of freedom at the end of the second phase (the instant of hitting the ball) and a timing parameter t_{hit} that controls when the swing back phase is transitioning to the hitting phase. For this task we employ a variant of the motor primitives that allows to set non-zero end velocities (Kober et al, 2010a). We learn these 15 meta-parameters as a function of the state, which

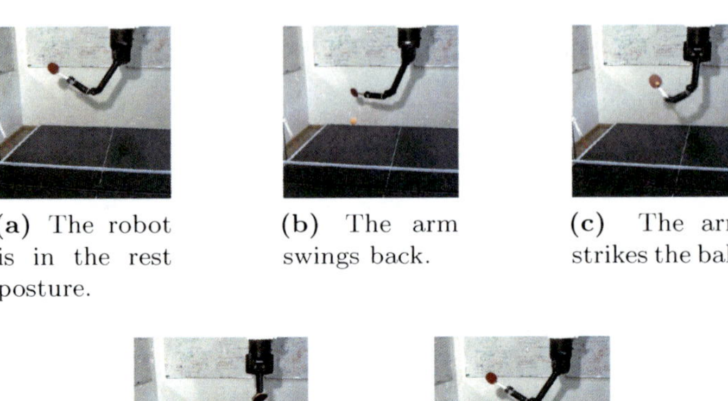

(a) The robot is in the rest posture.

(b) The arm swings back.

(c) The arm strikes the ball.

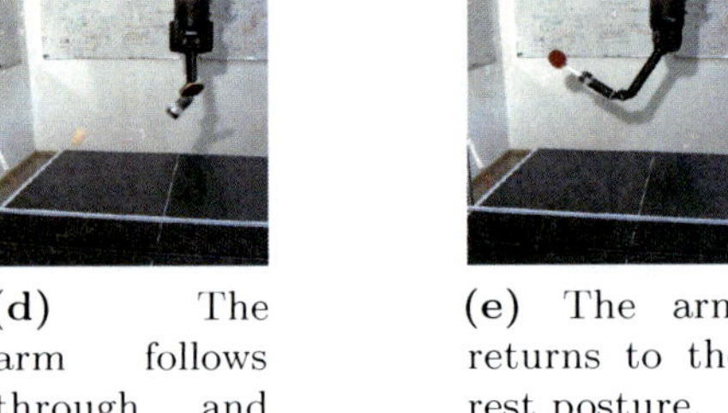

(d) The arm follows through and decelerates.

(e) The arm returns to the rest posture.

Fig. 5.11 This figure shows the phases of a table tennis stroke on the real Barrett WAM

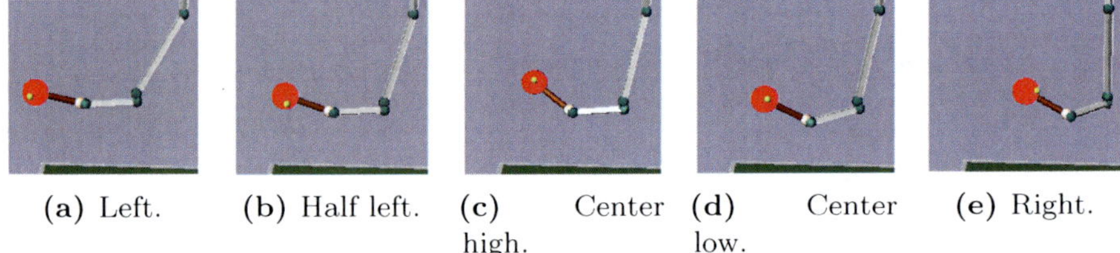

(a) Left.

(b) Half left.

(c) Center high.

(d) Center low.

(e) Right.

Fig. 5.12 This figure shows samples of the learned forehands. Note that this figure only illustrates the learned meta-parameter function in this context but cannot show timing (see Figure 5.13) and velocity and it requires a careful observer to note the important configuration differences resulting from the meta-parameters.

corresponds to the ball positions and velocities when it is directly over the net. We employed a Gaussian kernel and optimized the open kernel parameters according to typical values for the input and output beforehand. As cost function we employ

$$c = \sqrt{\sum_{i \in \{x,y,z\}} (b_i(t_{\mathrm{hit}}) - p_i(t_{\mathrm{hit}}))^2},$$

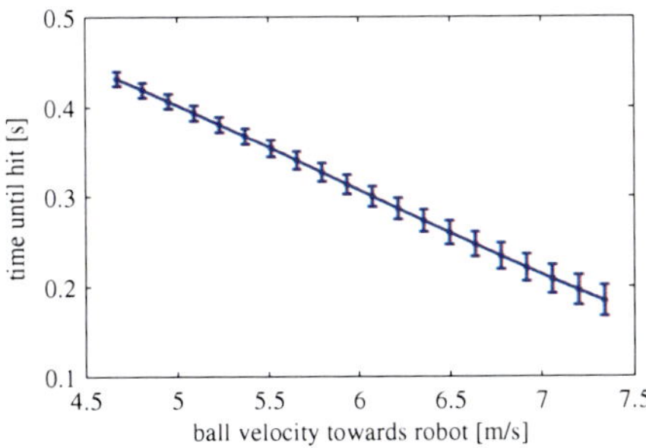

Fig. 5.13 This figure illustrates the effect of the velocity of the ball towards the robot on the time it has until the ball needs to be hit. The plot was generated by sweeping through the velocity component towards the robot, keeping the other position and velocity values fixed. The line is the mean of 100 sweeps drawn from the same ball distribution as used during the learning.

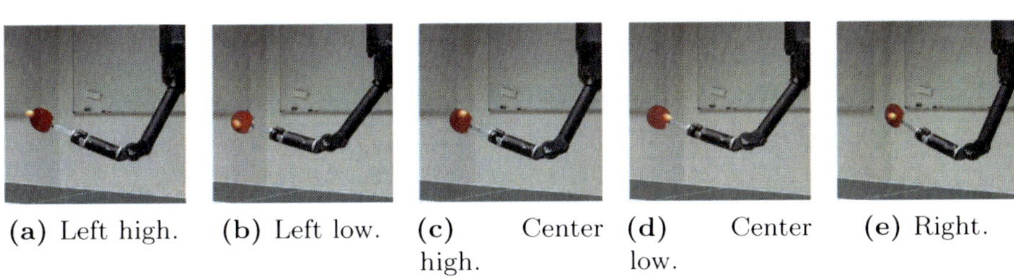

(a) Left high.   **(b)** Left low.   **(c)**   Center high.   **(d)**   Center low.   **(e)** Right.

Fig. 5.14 This figure shows samples of the learned forehands on the real robot

where $b_i (t_{\text{hit}})$ are the Cartesian positions of the ball and $p_i (t_{\text{hit}})$ are the Cartesian positions of the center of the paddle, both at the predicted hitting time t_{hit}. The policy is evaluated every 50 episodes with 25 ball launches picked randomly at the beginning of the learning. We initialize the behavior with five successful strokes observed from another player. After initializing the meta-parameter function with only these five initial examples, the robot misses approximately 95% of the balls as shown in Figure 5.15. Trials are only used to update the policy if the robot has successfully hit the ball as they did not significantly improve the learning performance and in order to keep the calculation sufficiently fast. Figures 5.12 and 5.14 illustrate different positions of the ball the policy is capable of dealing with after the learning. Figure 5.13 illustrates the dependence of the timing parameter on the ball velocity towards the robot and Figure 5.15 illustrates the costs over all episodes. For the results in Figure 5.15, we have simulated the flight of the ball as a simple ballistic point mass and the bouncing behavior using a restitution constant for the velocities. The state is directly taken from the simulated ball data with some added Gaussian noise. In the real robot experiment (Figure 5.16), the ball is shot with a ball cannon. The position of the ball is determined by two pairs of stereo cameras and the velocity is obtained by numerical differentiation. In this second setting, the state information is a lot less reliable due to noise in the vision system and even the same observed state can lead to different outcomes due to unobserved spin.

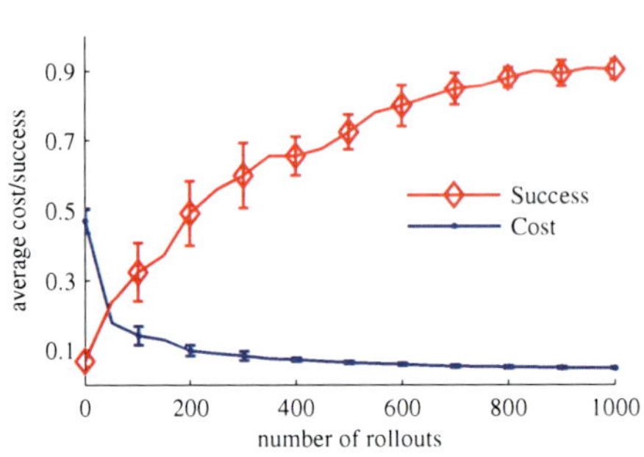

Fig. 5.15 This figure shows the cost function of the simulated table tennis task averaged over 10 runs with the error-bars indicating standard deviation. The red line represents the percentage of successful hits and the blue line the average cost. The number of rollouts includes the rollouts not used to update the policy. At the beginning the robot misses the ball 95% of the episodes and on average by $50\,cm$. At the end of the learning the robot hits almost all balls.

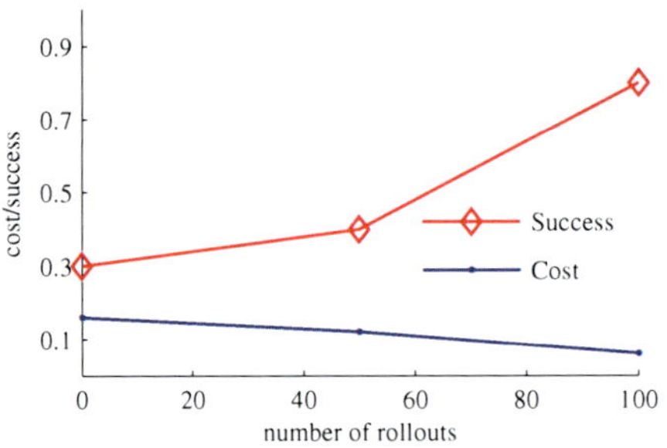

Fig. 5.16 This figure shows the cost function of the table tennis task on the real robot. The policy was learned entirely on the real robot. The red line represents the percentage of successful hits and the blue line the average cost. The number of rollouts includes the rollouts not used to update the policy. At the beginning the robot misses the ball 70% of the episodes and on average by $15\,cm$. At the end of the learning the robot hits 80% of the balls.

5.3.4 Active Learning of Ball Throwing

As an active learning setting, we chose a ball throwing task where the goal is to improve the throws while trying to perform well in a higher level game. For this scenario, it is important to balance learning of the individual actions by practicing them while at the same time, focusing on the overall performance in order to achieve the complete skill. Prominent examples are leisure time activities such as sports or motor skill games. For example, when playing darts with friends, you will neither always attempt the lowest risk action, nor always try to practice one particular throw, which will be valuable when mastered. Instead, you are likely to try plays with a reasonable level of risk and rely on safe throws in critical situations. This exploration is tightly woven into higher order dart games.

The higher level is modeled as a standard reinforcement learning problem with discrete states and actions. The lower level learning is done using CrKR. The higher level determines the target the robot is supposed to hit. The lower level has to learn how to hit this target. The transition probabilities of the higher level can be estimated from the learned meta-parameter function as explained in Section 5.3.4. We will discuss the rules of the game

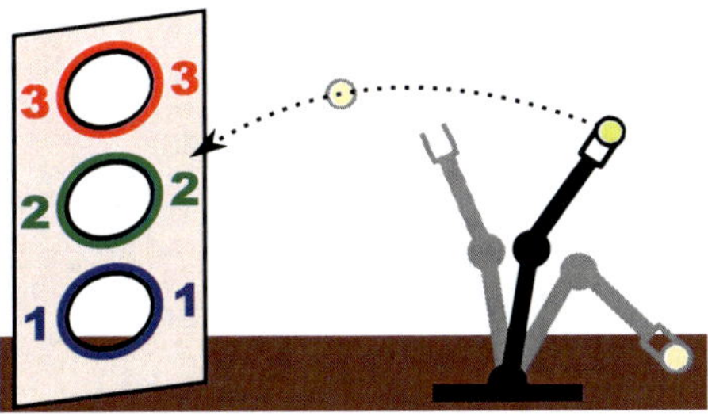

Fig. 5.17 This figure illustrates the side-stall game. The player throws the ball and if it lands in the target (illustrated by a wall with target holes) gets the number of points written next to it. Missing the targets is not punished, however, going over ten points leads to a loss of ten points.

in Section 5.3.4, a simulated experiment in Section 5.3.4, and the results of an evaluation with a real BioRob in Section 5.3.4.

Game Used for the Evaluations

The game is reminiscent of blackjack as the goal is to collect as many points as possible without going over a threshold. The player throws a ball at three targets. The three rewards of one, two, and three are assigned to one target each. The setup of the game is illustrated in Figure 5.17. If the ball lands in the target, the player receives the corresponding number of points. The player starts with zero points if he gets more than 10 points he "busts" and incurs a loss of -10. The player has the option to "stand" (i.e., stop throwing and collect the accumulated number of points) at all times. Missing all targets does not entail a cost.

Two-Level Learning Approach

Our framework considers a hierarchy of two levels: a strategy level and a behavior level. The strategy level determines the strategy for the high-level moves, here termed "behaviors", of the game. The behavior level deals with executing these behaviors in an optimal fashion. The strategy level chooses the next behavior, which is then executed by the behavior level. Upon completion of the behavior, the strategy level chooses the next behavior. The setup is illustrated in Figure 5.18.

We assume that the game has discrete states $s \in \mathbb{S}$ and discrete behaviors $b \in \mathbb{B}$. In the dart setting a behavior could be attempting to hit a specific field and the state could correspond to the current score. Given the current state, each behavior has an associated expected outcome $o \in \mathbb{O}$. For example, the behavior "throw at target X" has the outcome "change score by X" as a result of hitting target X. The transition probabilities $\mathcal{P}_{so}^{b}$ of the strategy level would express how likely it is to hit a different field. The game can be modeled as an Markov decision process or MDP (Sutton and Barto, 1998), where the states consist of the number of accumulated points (zero to ten) and two additional game states ("bust" and "stand"). The behaviors correspond to attempting to throw at a specific target or to "stand" and are fixed beforehand. We

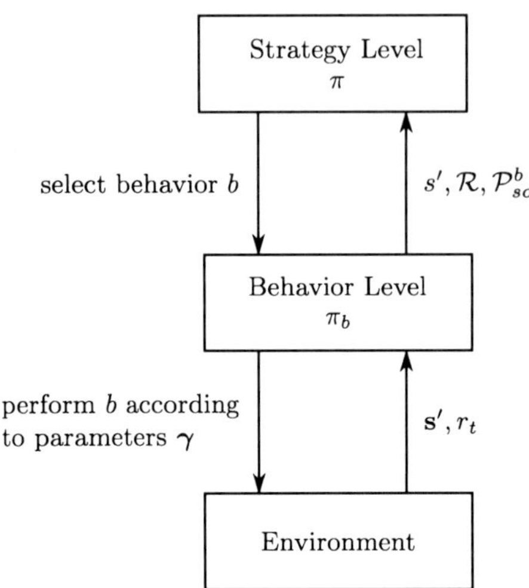

Fig. 5.18 This figure illustrates the setup of the roles of the different levels

assume to have an episodic game with a finite horizon, which can be expressed equivalently as an infinite horizon problem where we define an absorbing terminal state in which all actions receive an immediate reward of 0.

On the behavior level, we augment the state space with continuous states that describe the robot and the environment to form the combined state space **s**. This state space could, for example, include the position and velocity of the arm, the position of the targets as well as the current score. The actions are considered to be continuous and could, for example, be the accelerations of the arm. As the strategy level has to wait until the behavior is completed, the behaviors need to be of episodic nature as well. We have a single motor primitive representing the three behaviors of aiming at the three targets. Hitting the desired target is learned using CrKR. We employ Policy Iteration (Sutton and Barto, 1998) to learn on the strategy level.

The rewards for the strategy learning are fixed by the rules of the game. The possible states and behaviors also result from the way the game is played. The missing piece for the strategy learning is the transition probabilities $\mathcal{P}^b_{so}$. The behavior learning by CrKR associates each behavior with a variance. Each of these behaviors correspond to an expected change in state, the outcome o. For example "aim at 2" corresponds to "increase score by 2". However, the meta-parameter function does not explicitly include information regarding what happens if the expected change in state is not achieved. We assume that there is a discrete set of outcomes $o \in \mathbb{O}$ (i.e., change in state) for all behaviors b for a certain state s. For example in this game hitting each target, and missing, is associated with either increasing the player's score, winning or to bust (i.e., going over ten). With the meta-parameter function, we can calculate the overlaps of the ranges of possible meta-parameters for the different behaviors. These overlaps can then be used to determine how likely it is to end up with a change of state associated with a behavior different from

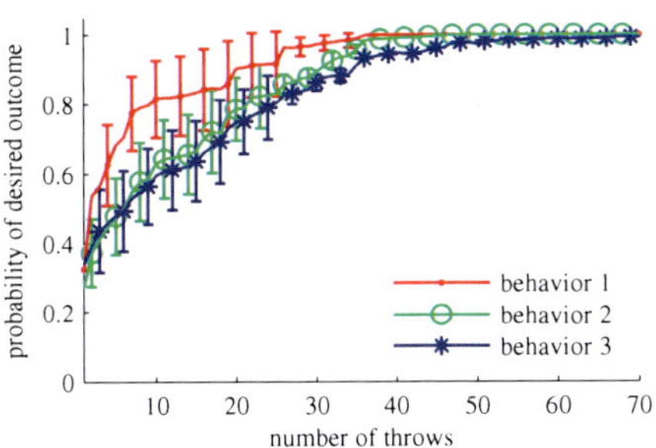

Fig. 5.19 This figure illustrates the transition probabilities of the three behaviors to their associated outcome in simulation. For example, the red line indicates the probability of gaining one point when throwing at target 1. After approximately 50 throws the player has improved his accuracy level such that he always hits the desired target. The plots are averaged over 10 runs with the error-bars indicating standard deviations.

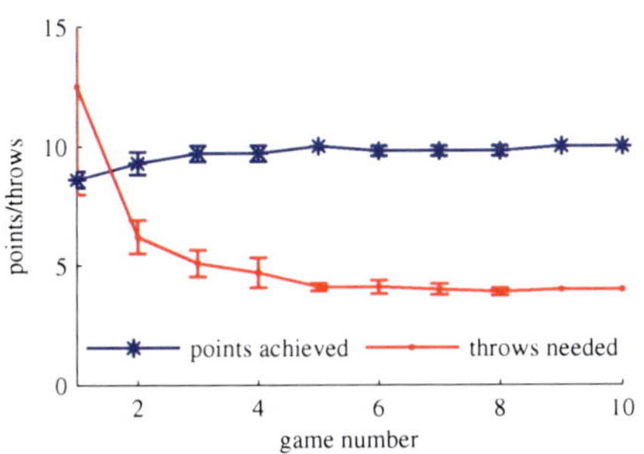

Fig. 5.20 This figure illustrates the improvement of the player over the number of games in simulation. Due to the large penalty for busting the framework always uses a safe strategy. Already after five completed games the player reaches almost always the maximum possible score of 10. As the number of throws is not punished there are initially many throws that miss the target. After 7 games the number of throws has converged to 4, which is the minimum required number. The plots are averaged over 10 runs with the error-bars indicating standard deviations.

the desired one. This approach relies on the assumption that we know for each behavior the associated range of meta-parameters and their likelihood.

The meta-parameters are drawn according to a normal distribution, thus the overlap has to be weighted accordingly. The probability of the outcome o when performing behavior b can be calculated as follows:

$$\mathcal{P}^b_{so} = \int p^b(\gamma) \frac{p^o(\gamma)}{\sum_{k \in \mathbb{O}} p^k(\gamma)} d\gamma,$$

where γ is the meta-parameters, $p^b(\gamma)$ is the probability of picking the meta-parameter γ when performing behavior b, $p^o(\gamma)$ is the probability of picking the meta-parameter γ when performing the action associated to the considered outcome o, and $\sum_{k \in \mathbb{O}} p^k(\gamma)$ is the normalizing factor. This scenario has first been treated in (Kober and Peters, 2011a).

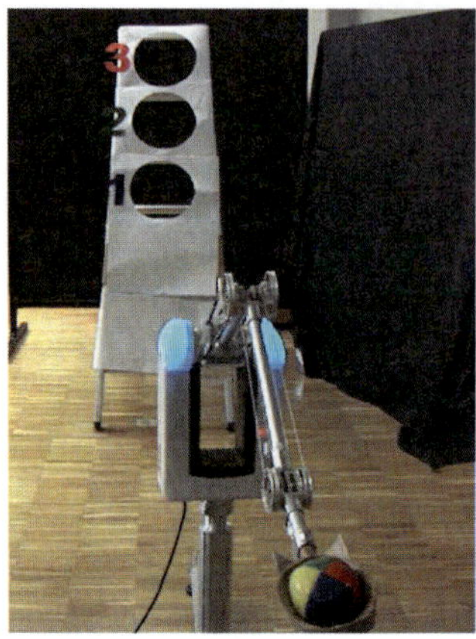

Fig. 5.21 This figure illustrates the setup of the robot evaluation

Evaluation in Simulation

We first evaluated our approach using a MATLAB based simulation. The throw is modeled as a two dimensional ballistic flight of a point mass. The targets correspond to segments of the ground line. The meta-parameters are the initial horizontal and vertical velocities of the ball. The meta-parameters used to initialize the learning make the ball drop in front of the first target. The cost function for the behavior level is

$$c = \sum_{i \in \{x,z\}} \dot{b}_i^2 + (b_x - s_x)^2 ,$$

where $\dot{b}_i$ are the initial velocities, b_x is the impact position and s_x the desired impact position. The state corresponds to the three targets and is determined by the higher level. Figure 5.19 illustrates how the player learns to throw more accurately while playing. Figure 5.20 illustrates how learning to perform the lower level actions more reliably enables the player to perform better in the game.

Evaluation on a Real BioRob

We employ a BioRob to throw balls in a catapult like fashion. The arm is approximately $0.75m$ long, and it can reach $1.55m$ above the ground. The targets are located at a distance of $2.5m$ from the robot at a height of $0.9m$, $1.2m$, and $1.5m$ respectively. The ball is placed in a funnel-shaped receptacle. In this setup, the initial horizontal and vertical velocities of the ball cannot directly be set. Instead, the meta-parameters are defined as the duration and amount of acceleration for two joints that are in the throwing plane. The robot starts in a fixed initial position, accelerates the two joints according to the meta-parameter indicating the magnitude, and accelerates in the opposite

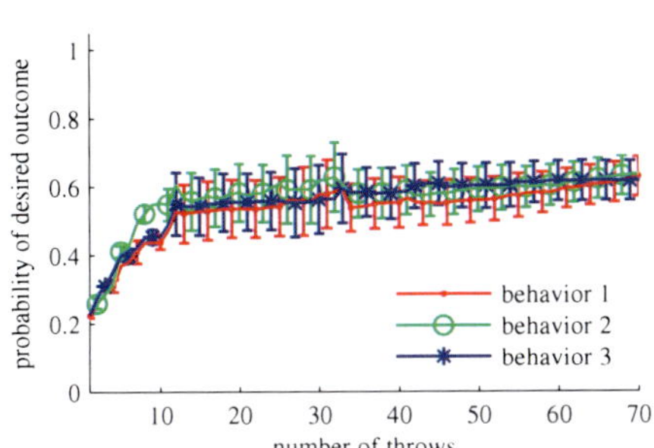

Fig. 5.22 This figure illustrates the transition probabilities of the three behaviors to their associated outcome like in Figure 5.19. The skill improves a lot in the first 15 throws after that the improvement levels of. Initially behavior 2, associated with target 2 (which lies in the center) is most likely to succeed. The success rate of 60% corresponds to the level of reproducibility of our setup. The framework manages to handle this large uncertainty by choosing to "stand" early on. The plots are averaged over 4 runs with the error-bars indicating standard deviations.

direction after the time determined by the other meta-parameter in order to break. Finally the robot returns to the initial position. See Figure 5.23 for an illustration of one throwing motion. The state corresponds to the three targets and is determined by the higher level. The outcome of the throw is observed by a vision system.

Executing the throw with identical parameters will only land at the same target in approximately 60% of the throws, due to the high velocities involved and small differences in putting the ball in the holder. Thus, the algorithm has to deal with large uncertainties. The cost function for the behavior level is

$$c = \sum_{i \in \{1,2\}} \ddot{\theta}_i^2 + t_{\text{acc}}^2 + (b_x - s_x)^2 \,,$$

where $\ddot{\theta}_i$ is the acceleration magnitude, t_{acc} the acceleration duration, b_x is the impact position and s_x the desired impact position. The setup makes it intentionally hard to hit target 3. The target can only be hit with a very restricted set of parameters. For targets 1 and 2 increasing the amount of acceleration or the duration will result in a higher hit. Target 3 is at the limit where higher accelerations or longer durations will lead to a throw in a downward direction with a high velocity.

The typical behavior of one complete experiment is as follows: At the beginning the robot explores in a very large area and stands as soon as it reaches a score of 8, 9, or 10. Due to the large punishment it is not willing to attempt to throw at 1 or 2 while having a large uncertainty, and, thus, a high chance of busting. Later on, it has learned that attempting to throw at 2 has a very low chance of ending up in 3 and hence will attempt to throw 2 points if the current score is 8. We setup the policy iteration to favor behaviors with a higher number, if the values of the behaviors are identical. The first throws of a round will often be aimed at 3, even if the probability of hitting target 2 using this action is actually higher than hitting the associated target 3. Until

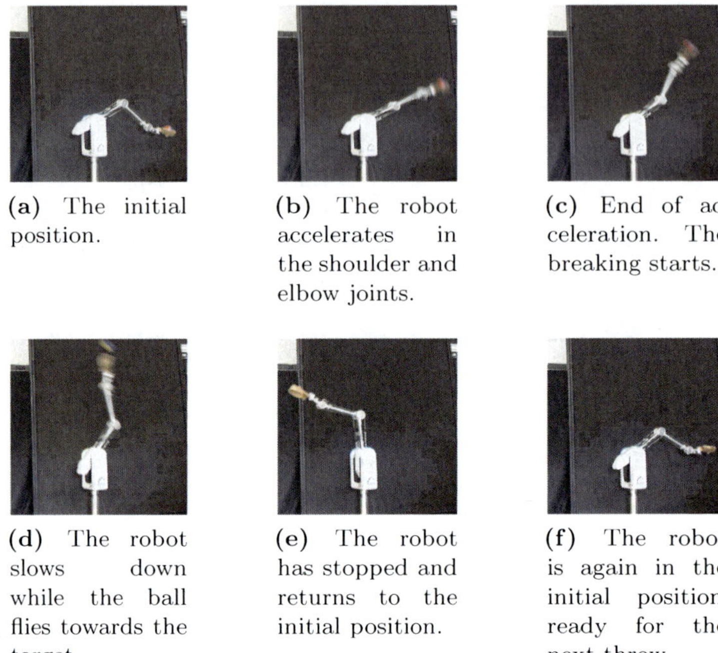

(a) The initial position.

(b) The robot accelerates in the shoulder and elbow joints.

(c) End of acceleration. The breaking starts.

(d) The robot slows down while the ball flies towards the target.

(e) The robot has stopped and returns to the initial position.

(f) The robot is again in the initial position, ready for the next throw.

Fig. 5.23 These frames illustrate one throwing motion with the BioRob

8 or more points have been accumulated, action 3 is safe (i.e., cannot lead to busting), does not entrain a punishment if missing or hitting a lower target, and has a large learning potential. Figure 5.22 illustrates how the robot learns to throw more accurately within the physical limits of the system.

5.4 Conclusion and Future Work

In this chapter, we have studied the problem of meta-parameter learning for motor primitives. It is an essential step towards applying motor primitives for learning complex motor skills in robotics more flexibly. We have discussed an appropriate reinforcement learning algorithm for mapping situations to meta-parameters.

We show that the necessary mapping from situation to meta-parameter can be learned using a Cost-regularized Kernel Regression (CrKR) while the parameters of the motor primitive can still be acquired through traditional approaches. The predictive variance of CrKR is used for exploration in on-policy meta-parameter reinforcement learning. We compare the resulting algorithm in a toy scenario to a policy gradient algorithm with a well-tuned policy representation and the reward-weighted regression. We show that our

CrKR algorithm can significantly outperform these preceding methods. We also illustrate the advantages of our reinforcement learning approach over supervised learning approaches in this setting. To demonstrate the system in a complex scenario, we have chosen the *Around the Clock* dart throwing game, table tennis, and ball throwing implemented both on simulated and real robots. In these scenarios we show that our approach performs well in a wide variety of settings, i.e. on four different real robots (namely a Barrett WAM, a BioRob, the JST-ICORP/SARCOS CBi and a Kuka KR 6), with different cost functions (both with and without secondary objectives), and with different policies in conjunction with their associated meta-parameters.

In the ball throwing task, we have discussed first steps towards a supervisory layer that deals with sequencing different motor primitives. This supervisory layer is learned by an hierarchical reinforcement learning approach (Huber and Grupen, 1998; Barto and Mahadevan, 2003). In this framework, the motor primitives with meta-parameter functions could also be seen as robotics counterpart of options (McGovern and Barto, 2001) or macro-actions (McGovern et al, 1997). The presented approach needs to be extended to deal with different actions that do not share the same underlying parametrization. For example in a table tennis task the supervisory layer would decide between a forehand motor primitive and a backhand motor primitive, the spatial meta-parameter and the timing of the motor primitive would be adapted according to the incoming ball, and the motor primitive would generate the trajectory. Future work will require to automatically detect which parameters can serve as meta-parameters as well as to discovering new motor primitives.

6

Learning Prioritized Control of Motor Primitives

Summary. Many tasks in robotics can be decomposed into sub-tasks that are performed simultaneously. In many cases, these sub-tasks cannot all be achieved jointly and a prioritization of such sub-tasks is required to resolve this issue. In this chapter, we discuss a novel learning approach that allows to learn a prioritized control law built on a set of sub-tasks represented by motor primitives. The primitives are executed simultaneously but have different priorities. Primitives of higher priority can override the commands of the conflicting lower priority ones. The dominance structure of these primitives has a significant impact on the performance of the prioritized control law. We evaluate the proposed approach with a ball bouncing task on a Barrett WAM.

6.1 Introduction

When learning a new skill, it is often easier to practice the required sub-tasks separately and later on combine them to perform the task – instead of attempting to learn the complete skill as a whole. For example, in sports sub-tasks can often be trained separately. Individual skills required in the sport are trained in isolation to improve the overall performance, e.g., in volleyball a serve can be trained without playing the whole game.

Sub-tasks often have to be performed simultaneously and it is not always possible to completely fulfill all at once. Hence, the sub-tasks need to be prioritized. An intuitive example for this kind of prioritizing sub-tasks happens during a volleyball game: a player considers hitting the ball (and hence avoiding it touching the ground and his team loosing a point) more important than locating a team mate and playing the ball precisely to him. The player will attempt to fulfill both sub-tasks. If this is not possible it is often better to "safe" the ball with a high hit and hope that another player recovers it rather than immediately loosing a point.

J. Kober and J. Peters, *Learning Motor Skills*,
Springer Tracts in Advanced Robotics 97,
DOI: 10.1007/978-3-319-03194-1_6, © Springer International Publishing Switzerland 2014

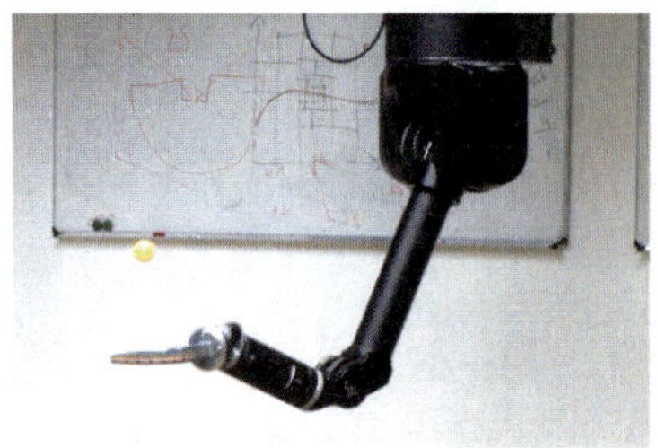

Fig. 6.1 This figure illustrates the ball-bouncing task on a Barrett WAM. The goal is to keep the ball bouncing on the racket.

In this chapter, we learn different sub-tasks that are represented by motor primitives that combined can perform a more complicated task. For doing so, we will stack controls corresponding to different primitives that represent movements in task space. These primitives are assigned different priorities and the motor commands corresponding to primitives with higher priorities can override the motor commands of lower priority ones. The proposed approach is outlined in Section 6.1.1 and further developed in Section 6.2. We evaluate our approach with a ball-bouncing task (see Figure 6.1 and Section 6.3).

As the sub-tasks describe the movements in task space, we have to learn a control that is mapping to the robot joint space. Unfortunately, this mapping is not a well-defined function for many robots. For example, if the considered task space has less degrees of freedom than the robot, multiple solutions are possible. This redundancy can be resolved by introducing a null-space control, i.e., a behavior that operates on the redundant degrees of freedom. Such a null-space control can for example pull the robot towards a rest posture (Peters and Schaal, 2008c), prevent getting close to joint limits (Chaumette and Marchand, 2001), avoid obstacles (Khatib, 1986) or singularities (Yoshikawa, 1985). Computing the task space control often corresponds to an optimization problem, that can for example be solved by a gradient based approach. A well known approach is the pseudo-inverse solution (Khatib, 1986; Peters and Schaal, 2008c). An alternative is to learn an operational space control law that implicitly includes the null-space behavior (Peters and Schaal, 2008a). Once learned, it corresponds to a unique mapping from desired actions in operational space to required actions in joint space.

The problem studied in this chapter is related to hierarchical control problems as discussed in (Findeisen et al, 1980). Using prioritized primitives in classical control has been explored in (Sentis and Khatib, 2005) by using analytical projections into the null-space. In this chapter, we propose a learning approach that does not require complete knowledge of the system, the constraints, and the task. In the reinforcement learning community, the compositions of options (i.e., concurrent options), which is related to the concurrent execution of primitives, has been studied by Precup et al (1998). Learning null-space control has been explored in (Towell et al, 2010). In contrast, we do not attempt to recover the implicit null-space policy but build a hierarchical operational space control law from user demonstrated primitives.

6.1.1 Proposed Approach

Based on the observation that many tasks can be described as a superposition of sub-tasks, we want to have a set of controls that can be executed simultaneously. As a representation for the sub-tasks, we chose the dynamical systems motor primitives, which are discussed in more detail in Chapter 3. Such primitives are well suited as representation for the sub-tasks as they ensure the stability of the movement generation. They are invariant under transformations of initial position and velocity, the final position and velocity, the duration as well as the movement amplitude.

In this chapter, these primitives are described in different task spaces, e.g., in the form

$$\ddot{\mathbf{x}}_i = \pi_i(\mathbf{x}_i, \dot{\mathbf{x}}_i, z)$$

where z denotes a share canonical system while $\mathbf{x}_i$ are positions in task-space i. For example, if we have a primitive "move end-effector up and down" its task space would correspond to the Cartesian position indicating the height (as well as the corresponding velocities and accelerations) but not include the sideways movement or the orientation of the end-effector. The dynamical systems motor primitives are well suited to represent different kinds of vertical movements starting and ending at various states and of different duration.

These primitives are prioritized such that

$$i \succeq i - 1,$$

which reads a "task i dominates task $i - 1$". If both sub-tasks can be fulfilled at the same time, our system will do so – but if this should not be possible, sub-task i will be fulfilled at the expense of sub-task $i - 1$. We attempt to reproduce a complex task that consists of several sub-tasks, represented by motor primitives,

$$\{\pi_1, \pi_2, \ldots, \pi_N\}$$

that are concurrently executed at the same time following the prioritization scheme

$$N \succeq N - 1 \succeq \cdots \succeq 2 \succeq 1.$$

This approach requires a prioritized control law that composes the motor command out of the primitives π_i, i.e.,

$$\mathbf{u} = \mathbf{f}(\pi_1, \pi_2, \ldots, \pi_N, \mathbf{q}, \dot{\mathbf{q}})$$

where $\mathbf{q}, \dot{\mathbf{q}}$ are the joint position and joint velocity, $\mathbf{u}$ are the generated motor commands (torques or accelerations).

We try to acquire the prioritized control law in three steps, which we will illustrate with the ball-bouncing task:

1. We observe $\ddot{\mathbf{x}}_i(t), \dot{\mathbf{x}}_i(t), \mathbf{x}_i(t)$ individually for each of the primitives that will be used for the task. For the ball-bouncing example, we may have the

following sub-tasks: "move under the ball", "hit the ball", and "change racket orientation". The training data is collected by executing only one primitive at a time without considering the global strategy, e.g., for the "change racket orientation" primitive by keeping the position of the racket fixed and only changing its orientation without a ball being present. This training data is used to acquire the task by imitation learning under the assumption that these tasks did not need to overrule each other in the demonstration (Sect. 6.2).

2. We enumerate all possible dominance structures and learn a prioritized control law for each dominance list that fusions the motor primitives. For the three ball-bouncing primitives there are six possible orders, as listed in Table 6.1.

3. We choose the most successful of these approaches. The activation and adaptation of the different primitives is handled by a strategy layer (Section 6.3.2). In the ball-bouncing task, we evaluate how long each of the prioritized control laws keeps the ball in the air and pick the best performing one (Section 6.3.3).

Clearly, enumerating all possible dominance structures only works for small systems (as the number of possibilities grows with $n!$, i.e., exponentially fast).

6.2   Learning the Prioritized Control Law

By learning the prioritized control, we want to obtain a control law

$$\mathbf{u} = \ddot{\mathbf{q}} = \mathbf{f}(\pi_1, \pi_2, \ldots, \pi_N, \mathbf{q}, \dot{\mathbf{q}}),$$

i.e., we want to obtain the required control $\mathbf{u}$ that executes the primitives $\pi_1, \pi_2, \ldots, \pi_N$. Here, the controls correspond to the joint accelerations $\ddot{\mathbf{q}}$. The required joint accelerations not only depend on the primitives but also on the current state of the robot, i.e., the joint positions $\mathbf{q}$ and joint velocities $\dot{\mathbf{q}}$. Any control law can be represented locally as a linear control law. In our setting, these linear control laws can be represented as

$$\mathbf{u} = \begin{bmatrix} \ddot{\mathbf{x}}_i \\ \dot{\mathbf{q}} \\ \mathbf{q} \end{bmatrix}^{\mathrm{T}} \boldsymbol{\theta} = \boldsymbol{\phi}^{\mathrm{T}} \boldsymbol{\theta},$$

where $\boldsymbol{\theta}$ are the parameters we want to learn and $\boldsymbol{\phi} = \begin{bmatrix} \ddot{\mathbf{x}}_i & \dot{\mathbf{q}} & \mathbf{q} \end{bmatrix}$ acts as features. Often the actions of the primitive $\ddot{\mathbf{x}}_i$ can be achieved in multiple different ways due to the redundancies in the robot degrees of freedom. To ensure consistency, a null-space control is introduced. The null-space control can, for example, be defined to pull the robot towards a rest posture $\mathbf{q}_0$, resulting in the null-space control

$$\mathbf{u}_0 = -\mathbf{K}_D \dot{\mathbf{q}} - \mathbf{K}_P \left(\mathbf{q} - \mathbf{q}_0 \right),$$

where $\mathbf{K}_D$ and $\mathbf{K}_P$ are gains for the velocities and positions respectively.

To learn the prioritized control law, we try to generalize the learning operational space control approach from (Peters and Schaal, 2008a) to a hierarchical control approach (Sentis and Khatib, 2005; Peters and Schaal, 2008c).

6.2.1 *Single Primitive Control Law*

A straightforward approach to learn the motor commands $\mathbf{u}$, represented by the linear model $\mathbf{u} = \boldsymbol{\phi}^{\mathrm{T}}\boldsymbol{\theta}$, is using linear regression. This approach minimizes the squared error

$$\mathrm{E}^2 = \sum_{t=1}^{T} \left(\mathbf{u}_t^{\mathrm{ref}} - \boldsymbol{\phi}_t^{\mathrm{T}}\boldsymbol{\theta}\right)^2$$

between the demonstrated control of the primitive u_t^{ref} and the recovered linear policy $\mathbf{u}_t = \boldsymbol{\phi}_t^{\mathrm{T}}\boldsymbol{\theta}$, where T is the number of samples. The parameters minimizing this error are

$$\boldsymbol{\theta} = \left(\boldsymbol{\Phi}^T\boldsymbol{\Phi} + \lambda\mathbf{I}\right)^{-1}\boldsymbol{\Phi}^{\mathrm{T}}\mathbf{U}, \tag{6.1}$$

with $\boldsymbol{\Phi}$ and $\mathbf{U}$ containing the values of the demonstrated ϕ and $\mathbf{u}$ for all time-steps t respectively, and a ridge factor λ. If the task space and the joint-space coincide, the controls $\mathbf{u} = \ddot{\mathbf{q}}$ are identical to the action of the primitive $\ddot{\mathbf{x}}_i$. We also know that *locally* any control law that can be learned from data is a viable control law (Peters and Schaal, 2008a). The error with respect to the training data is minimized, *however*, if the training data is not consistent, the plain linear regression will average the motor commands, which is unlikely to fulfill the actions of the primitive.

In order to enforce consistency, the learning approach has to resolve the redundancy and incorporate the null-space control. We can achieve this by using the program

$$\min_{\mathbf{u}} J = (\mathbf{u} - \mathbf{u}_0)^{\mathrm{T}}\mathbf{N}(\mathbf{u} - \mathbf{u}_0) \tag{6.2}$$

$$\mathrm{s.t.}\ \ddot{\mathbf{x}} = \pi(\mathbf{x}, \dot{\mathbf{x}}, z)$$

as discussed in (Peters and Schaal, 2008c). Here the cost J is defined as the weighted squared difference of the control $\mathbf{u}$ and the null-space control $\mathbf{u}_0$, where the metric $\mathbf{N}$ is a positive semi-definite matrix. The idea is to find controls $\mathbf{u}$ that are as close as possible to the null-space control $\mathbf{u}_0$ while still fulfilling the constraints of the primitive π. This program can also be solved as discussed in (Peters and Schaal, 2008a). Briefly speaking, the regression in Equation (6.1) can be made consistent by weighting down the error by weights w_t and hence obtaining

$$\boldsymbol{\theta} = \left(\boldsymbol{\Phi}^{\mathrm{T}}\mathbf{W}\boldsymbol{\Phi} + \lambda\mathbf{I}\right)^{-1}\boldsymbol{\Phi}^{\mathrm{T}}\mathbf{W}\mathbf{U}$$

Algorithm 6.1 Learning the Prioritized Control Law

---

define null-space control $\mathbf{u}_0$, metric $\mathbf{N}$, scaling factor α

collect controls $\mathbf{u}_{i,t}$ and features $\boldsymbol{\phi}_{i,t}$ for all primitives $i \in \{1, \ldots, N\}$ and all time-steps $t \in \{1, \ldots, T\}$ separately

for primitives $i = 1 \ldots N$ (N: highest priority) **do**

    for time-steps $t = 1 \ldots T$ **do**

        calculate offset controls
$$\hat{\mathbf{u}}_{i,t} = \mathbf{u}_{i,t} - \sum_{j=1}^{i-1} \boldsymbol{\phi}_{i,t}^{\mathrm{T}} \boldsymbol{\theta}_j - \mathbf{u}_{0,t}$$

        calculate weights $\hat{w}_{i,t} = \exp\left(-\alpha \hat{\mathbf{u}}_{i,t}^{\mathrm{T}} \mathbf{N} \hat{\mathbf{u}}_{i,t}\right)$

    end for

    build control matrix $\hat{\mathbf{U}}_i$ containing $\hat{\mathbf{u}}_{i,1} \ldots \hat{\mathbf{u}}_{i,T}$

    build feature matrix $\boldsymbol{\Phi}_i$ containing $\boldsymbol{\phi}_{i,1} \ldots \boldsymbol{\phi}_{i,T}$

    build weight matrix $\hat{\mathbf{W}}_i = \mathrm{diag}(\hat{w}_{i,1}, \ldots, \hat{w}_{i,T})$

    calculate parameters
$$\boldsymbol{\theta}_i = \left(\boldsymbol{\Phi}_i^{\mathrm{T}} \hat{\mathbf{W}}_i \boldsymbol{\Phi}_i + \lambda \mathbf{I}\right)^{-1} \boldsymbol{\Phi}_i^{\mathrm{T}} \hat{\mathbf{W}}_i \hat{\mathbf{U}}_i$$

    end for

end for

---

with $\mathbf{W} = \mathrm{diag}(w_1, \ldots, w_{Tn})$ for T samples. This approach works well for linear models and can be gotten to work with multiple locally linear control laws. Nevertheless, it maximizes a reward instead of minimizing a cost. The cost J can be transformed into weights w_t by passing it through an exponential function
$$w_t = \exp\left(-\alpha \tilde{\mathbf{u}}_t^{\mathrm{T}} \mathbf{N} \tilde{\mathbf{u}}_t\right),$$
where $\tilde{\mathbf{u}}_t = (\mathbf{u}_t - \mathbf{u}_0)$. The scaling factor α acts as a monotonic transformation that does not affect the optimal solution but can increase the efficiency of the learning algorithm.

Using the Woodbury formula (Welling, 2010) Equation (6.2.1) can be transformed into
$$\mathbf{u} = \boldsymbol{\phi}(x)^{\mathrm{T}} \boldsymbol{\Phi}^{\mathrm{T}} \left(\boldsymbol{\Phi}\boldsymbol{\Phi}^{\mathrm{T}} + \mathbf{W}_U\right)^{-1} \mathbf{U} \tag{6.3}$$
with $\mathbf{W}_U = \mathrm{diag}\left(\tilde{\mathbf{u}}_1^T \mathbf{N} \tilde{\mathbf{u}}_1, \ldots, \tilde{\mathbf{u}}_n^T \mathbf{N} \tilde{\mathbf{u}}_n\right)$. By introducing the kernels $\mathbf{k}(\mathbf{s}) = \boldsymbol{\phi}(\mathbf{s})^{\mathrm{T}} \boldsymbol{\Phi}^{\mathrm{T}}$ and $\mathbf{K} = \boldsymbol{\Phi}\boldsymbol{\Phi}^{\mathrm{T}}$ we obtain
$$\mathbf{u} = \mathbf{k}(\mathbf{s})^{\mathrm{T}} \left(\mathbf{K} + \mathbf{W}_U\right)^{-1} \mathbf{U},$$
which is related to the kernel regression (Bishop, 2006) and corresponds to the Cost-regularized Kernel Regression introduced in Chapter 5. This kernelized form of Equation (6.3) overcomes the limitations of the linear model at a cost of higher computational complexity.

6.2.2 Prioritized Primitives Control Law

In the previous section, we have described how the control law for a single primitive can be learned. To generalize this approach to multiple primitives with different priorities, we want a control law that always fulfills the primitive with the highest priority and follows the remaining primitives as much as possible according to their place in the hierarchy. Our idea is to represent the higher priority control laws as correction term with respect to the lower priority primitives. The control of the primitive with the lowest priority is learned first. This control is subsequently considered to be a baseline and the primitives of higher priority only learn the difference to this baseline control. The change between the motor commands resulting from primitives of lower priority is minimized. The approach is reminiscent of online passive-aggressive algorithms (Crammer et al, 2006). Hence, control laws of higher priority primitives only learn the offset between their desired behavior and the behavior of the lower priority primitives. This structure allows them to override the actions of the primitives of lesser priority and, therefore, add more detailed control in the regions of the state space they are concerned with. The combined control of all primitives is

$$\mathbf{u} = \mathbf{u}_0 + \sum_{n=1}^{N} \Delta\mathbf{u}_n,$$

where $\mathbf{u}_0$ is the null-space control and $\Delta\mathbf{u}_n$ are the offset controls of the N primitives.

Such control laws can be expressed by changing the program in Equation (6.2) to

$$\min_{\mathbf{u}_i} J = \left(\mathbf{u}_i - \sum_{j=1}^{i-1} \Delta\mathbf{u}_j - \mathbf{u}_0 \right)^{\mathrm{T}} \mathbf{N} \left(\mathbf{u}_i - \sum_{j=1}^{i-1} \Delta\mathbf{u}_j - \mathbf{u}_0 \right)$$

$$\text{s.t. } \ddot{\mathbf{x}}_i = \pi_i(\mathbf{x}_i, \dot{\mathbf{x}}_i, z),$$

where the primitives need to be learned in the increasing order of their priority, the primitive with the lowest priority is learned first, the primitive with the highest priority is learned last. The regression in Equation (6.2.1) changes to

$$\boldsymbol{\theta}_i = \left(\boldsymbol{\Phi}_i^{\mathrm{T}} \hat{\mathbf{W}}_i \boldsymbol{\Phi}_i + \lambda\mathbf{I} \right)^{-1} \boldsymbol{\Phi}_i^{\mathrm{T}} \hat{\mathbf{W}}_i \hat{\mathbf{U}}_i,$$

where $\hat{\mathbf{U}}_i$ contains the offset controls $\hat{\mathbf{u}}_{i,t} = \mathbf{u}_{i,t} - \sum_{j=1}^{i-1} \Delta\mathbf{u}_{j,t} - \mathbf{u}_{0,t}$ for all time-steps t, where $\Delta\mathbf{u}_{j,t} = \boldsymbol{\phi}_{i,t}^{\mathrm{T}} \boldsymbol{\theta}_j$. The weighting matrix $\hat{\mathbf{W}}_i$ now has the weights $\hat{w}_t = \exp\left(-\alpha\hat{\mathbf{u}}_{i,t}^{\mathrm{T}} \mathbf{N}\hat{\mathbf{u}}_{i,t} \right)$ on its diagonal and matrix $\hat{\mathbf{U}}_i$ contains offset controls $\hat{\mathbf{u}}_{i,t}$. The kernelized form of the prioritized control law can be obtained analogously. The complete approach is summarized in Algorithm 6.1.

(a) Exaggerated schematic drawing. The green arrows indicate velocities.

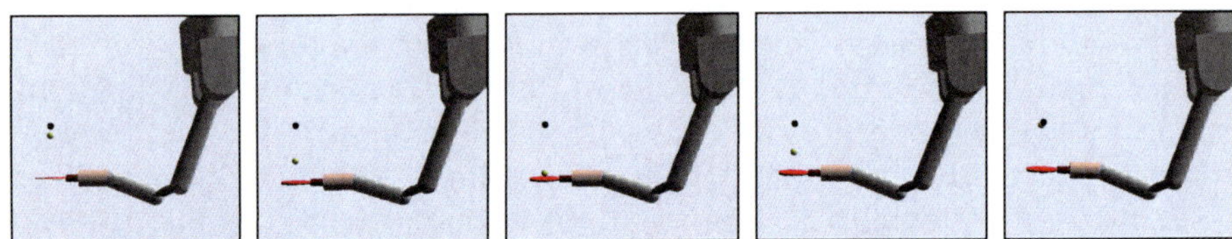

(b) Paddling movement for the simulated robot. The black ball represents the virtual target (see Section 6.3.2)

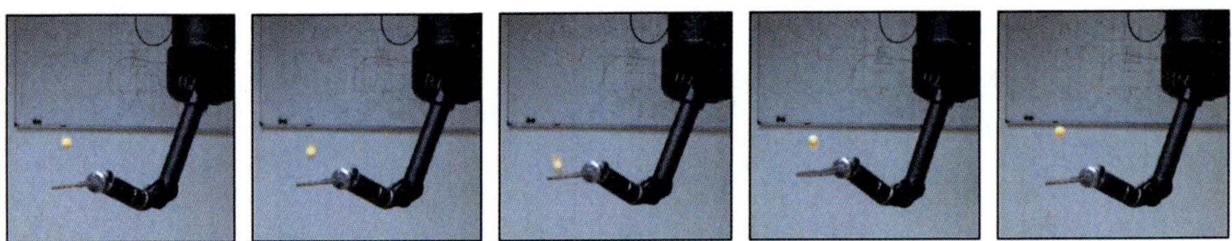

(c) Paddling movement for the real Barrett WAM.

Fig. 6.2 This figure illustrates a possible sequence of bouncing the ball on the racket in a schematic drawing, in simulation, and on the real robot

6.3   Evaluation: Ball-Bouncing

In order to evaluate the proposed prioritized control approach, we chose a ball bouncing task. We describe the task in Section 6.3.1, explain a possible higher level strategy in Section 6.3.2, and discuss how the proposed framework can be applied in Section 6.3.3.

6.3.1   Task Description

The goal of the task is to bounce a table tennis ball above a racket. The racket is held in the player's hand, or in our case attached to the end-effector of the robot. The ball is supposed to be kept bouncing on the racket. A possible movement is illustrated in Figure 6.2.

It is desirable to stabilize the bouncing movement to a strictly vertical bounce, hence, avoiding the need of the player to move a lot in space and, thus, leaving the work space of the robot. The hitting height is a trade-off between having more time until the next hit at the expense of the next hitting position possibly being further away. The task can be sub-dived into three

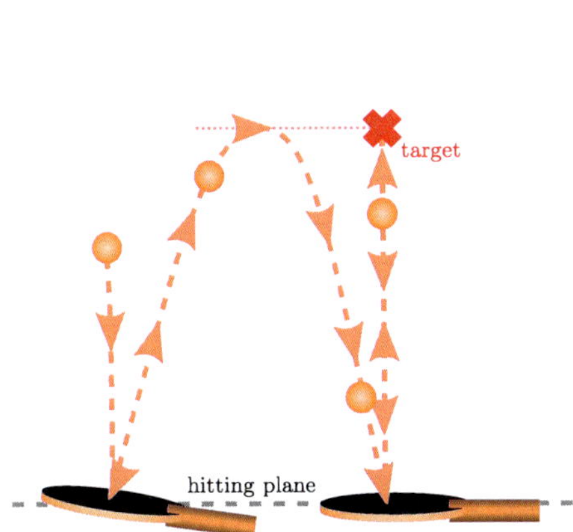

Fig. 6.3 This figure illustrates the employed strategy for bouncing the ball on the racket. The highest point of the ball trajectory is supposed to coincide with the red target. The racket is always hitting the ball in a fixed height, i.e., the hitting plane. The strategy is to play the ball in a way that the next intersection with the hitting plane is directly below the target and the maximum height of the ball trajectory corresponds to the height of the target. If the bounce works exactly as planned, the ball needs to be hit only once to return to a strictly vertical bouncing behavior.

intuitive primitives: hitting the ball upward, moving the racket under the ball before hitting, and changing the orientation of the racket to move the ball to a desired location. A possible strategy is outlined in the next section.

The ball is tracked using a stereo vision setup and its positions and velocities are estimated by a Kalman filter. To initialize the ball-bouncing task, the ball is thrown towards the racket.

6.3.2   Bouncing Strategy

The strategy employed to achieve the desired bouncing behavior is based on an imagined target that indicates the desired bouncing height. This target is above the default posture of the racket. The top point of the ball trajectory is supposed to hit this target, and the stable behavior should be a strictly vertical bounce. This behavior can be achieved by defining a hitting plane, i.e., a height at which the ball is always hit (which corresponds to the default posture of the racket). On this hitting plane, the ball is always hit the in a manner that the top point of its trajectory corresponds to the height of the target and the next intersection of the ball trajectory with the hitting plane is directly under the target. See Figure 6.3 for an illustration.

To achieve this desired ball behavior, the racket is always moved to the intersection point of the ball trajectory and the hitting plane. By choosing the hitting velocity and the orientation of the racket, the velocity and direction of the ball after being hit can be changed. The required hitting velocity and orientation are calculated using a model of the ball and the racket. The ball is modeled as a point mass that moves according to the ballistic flight equations. For the relatively low speeds and small distances air resistance is negligible. The contact with the racket is modeled as a reflection with a restitution factor.

Table 6.1 This table shows the suitability of the possible dominance structures (mean±std). The "hit the ball" primitive clearly is the dominant one, followed by the "move under the ball" primitive. The prioritized control laws work significantly better than a single model learned using the combined training data of the three primitives. Preliminary results on the real robot confirm this ordering.

Dominance Structure	Number of Hits	
	in Simulation	on Real Robot
single model	5.70 ± 0.73	1.10 ± 0.99
hit$\succeq$move$\succeq$orient	11.35 ± 2.16	2.30 ± 0.67
hit$\succeq$orient$\succeq$move	10.85 ± 1.46	1.70 ± 0.95
move$\succeq$hit$\succeq$orient	9.05 ± 0.76	1.40 ± 0.70
move$\succeq$orient$\succeq$hit	7.75 ± 1.48	1.40 ± 0.84
orient$\succeq$hit$\succeq$move	5.90 ± 0.85	1.30 ± 0.67
orient$\succeq$move$\succeq$hit	5.35 ± 0.49	1.30 ± 0.48

Using this strategy the ball can be brought back to a strictly vertical bouncing behavior with a single hit. However, this method requires the knowledge of the ball position and velocity, as well as a model of the ball behavior. An alternative strategy that stabilizes the behavior in a completely open loop behavior by employing a slightly concave paddle shape has been suggested in (Reist and D'Andrea, 2009). A method similar to the proposed strategy has been employed by Kulchenko (2011) and Müller et al (2011). Buehler et al (1994) proposed the mirror law for this task. The ball bouncing task has also be employed to study how humans stabilize a rhythmic task (Schaal et al, 1996).

6.3.3 *Learning Results*

As discussed in Section 6.3.1, the task can be described by three primitives: "move under the ball", "hit the ball", and "change racket orientation". Training data is collected in the relevant state space independently for each primitive. For doing so, the parameters corresponding to the other primitives are kept fixed and variants of the primitive are hence executed from various different starting positions. The primitive "move under the ball" corresponds to movements in the horizontal plane, the primitive "hit the ball" to up and down movements, and the primitive "change racket orientation" only changes the orientation of the end-effector. We collected 30 seconds of training data for each primitive, corresponding to approximately 60 bounces.

Having only three primitives allows it to enumerate all six possible dominance structures, to learn the corresponding prioritized control law, and to evaluate the controller. As intuitive quality measure we counted the number of bounces until the robot missed, either due to imprecise control or due to the ball being outside of the safely reachable work-space.

Table 6.1 illustrates the resulting dominance structures. The most relevant primitive is the "hit the ball" primitive, followed by the "move under the ball" primitive. In the table it is clearly visible that inverting the order of two neighboring primitives that are in the preferred dominance order always results in a lower number of hits. Compared to a single model, that was trained using the combined training data of the three primitives, all but two prioritized control laws work significantly better. The ordering may appear slightly counter-intuitive as moving under the ball seems to be the most important primitive in order to keep the ball in the air, allowing for later corrections. However, the robot has a fixed base position and the ball moves quickly out of the safely reachable work-space, resulting in a low number of hits. Additionally, the default position of the racket is almost vertical, hence covering a fairly large are of the horizontal plane resulting in robustness with respect to errors in this primitive.

6.4 Conclusion

In this chapter, we have presented a prioritized control learning approach that is based on the superposition of movement primitives. We have introduced a novel framework for learning prioritized control. The controls of the lower priority primitives are fulfilled as long as they lay in the null-space of the higher priority ones and get overridden otherwise. As representation for the primitives, we employ the dynamical systems motor primitives (Ijspeert et al, 2002b; Schaal et al, 2007), which yield controls in the form of desired accelerations. These primitives are executed separately to collect training data. Local linear models are trained using a weighted regression technique incorporating the various possible dominance structures. In the presented ball bouncing task, the movement is restricted to a space where the controls are approximately linear. Hence, a single linear model per primitive was sufficient. This limitation can be overcome by either considering multiple local linear models (Peters and Schaal, 2008a) or by kernelizing the weighted regression, as described in Section 6.2.1 and 6.2.2.

The dominance structure of the task was determined by testing all possible structures exhaustively. Intuitively, the lower priority primitives represent a global behavior and the high priority primitives represent specialized corrections, hence overriding the lower priority controls. In most cases, the resulting prioritized control works significantly better than a single layer one that was trained with the combined training data of all primitives. As illustrated by the evaluations, the dominance structure can have a significant influence on the global success of the prioritized control. Enumerating all possible dominance structures is factorial in the number of primitives and hence unfeasible in practice for more than four primitives. In this case, smarter search strategies are needed.

The success of the different dominance structures not only depends on the task but also on the employed strategy of activating and adapting the different primitives. An interesting area for future research could be to jointly learn the prioritized control and the strategy.

The presented approach has been evaluated both in simulation and on a real Barrett WAM and we have demonstrated that our novel approach can successfully learn a ball-bouncing task.

7

Conclusion

Summary. In this book, we have discussed reinforcement learning approaches for motor skills represented by motor primitives. In the next section, we provide an overview of the key contributions in this book and then we discuss possible directions for extending the presented research.

7.1 Contributions

The contributions of this book are for the state-of the-art in both machine learning and in robotics.

7.1.1 Algorithms

In this book, we have proposed a framework of policy search based on reward-weighted imitation. The resulting EM-inspired algorithms are applicable both to parametrized and non-parametric policies. The policy search algorithms presented in this book perform local optimization which results in fast convergence but also poses the risk of converging to bad local optima. For all the presented application scenarios, good initial policies are available, which mitigates the problem of bad local optima and renders the approaches applicable in practice.

In Section 1.2.1, we have discussed requirements for robotic reinforcement learning approaches, i.e., they should avoid damage to the robot, and should be fast, both in terms of convergence and computation time. Having a sample-efficient algorithm, only very few open parameters, and the ability to incorporate prior knowledge all are essential for fast convergence. In the following we will discuss how the proposed algorithms meet these requirements.

J. Kober and J. Peters, *Learning Motor Skills*,
Springer Tracts in Advanced Robotics 97,
DOI: 10.1007/978-3-319-03194-1_7, © Springer International Publishing Switzerland 2014

Policy learning by Weighting Exploration with the Returns (PoWER)

PoWER (Chapter 4) is an EM-inspired policy search algorithm relying on a parametrized policy and structured exploration in the parameter space. The algorithm is particularly suitable for motor primitives (Chapter 3). We introduce a framework of reward weighted imitation that yields several well known policy search algorithms: episodic REINFORCE (Williams, 1992), the policy gradient theorem (Sutton et al, 1999), episodic natural actor critic (Peters and Schaal, 2008b), a generalization of the reward-weighted regression (Peters and Schaal, 2008a).

PoWER is unlikely to damage the robot as it only explores locally and usually is initialized by an initial policy. Using importance sampling, the policy can be updated after each rollout, hence it is sample efficient. Prior knowledge can be incorporated both in the form of the initial policy and by defining the policy structure via the parametrization. The main open parameter is the exploration magnitude. This parameter only needs to be set once initially and can then be automatically adjusted during the learning process as discussed in Section 4.A.3. The policy structure is also an open parameter. The dynamical systems motor primitives essentially only have a single open parameter, i.e., the number of basis functions as an open parameter which corresponds to the amount of detail that the policy can capture. The bottleneck of the update calculation is a matrix inversion that can be avoided by making some additional independence assumptions that work well in conjunction with the dynamical systems motor primitives.

PoWER outperformed various other policy search approaches (i.e., finite difference gradients, episodic REINFORCE, 'vanilla' policy gradients with optimal baselines, episodic natural actor critic, and episodic reward-weighted regression) on benchmarks and robotic tasks. Our approach has inspired follow-up work in other contexts, for example (Vlassis et al, 2009; Kormushev et al, 2010). Theodorou et al (2010) have derived algorithms based on the path integral approach that are very similar to PoWER and have also been successfully employed for robotic tasks (Buchli et al, 2011; Kalakrishnan et al, 2011; Pastor et al, 2011; Stulp et al, 2011; Tamošiūnaitė et al, 2011).

Cost-regularized Kernel Regression (CrKR)

CrKR (Chapter 5) is an EM-inspired policy search algorithm employing a non-parametric policy representation. CrKR is based on a kernelized version of the episodic reward-weighted regression [Peters and Schaal, 2008a, Chapter 4]. Similar to Gaussian process regression, CrKR yields a predictive variance that can be employed to guide the exploration. In this book it is employed to generalize movements to new situations. This type of learning relies on a policy representation that has a number of meta-parameters that allow to generalize the movement globally while retaining the details of the motion. The motor primitives (Chapter 3) inherently have six modification parameters (the initial position $\mathbf{x}_1^0$, the initial velocity $\mathbf{x}_2^0$, the goal $\mathbf{g}$, the goal velocity $\dot{\mathbf{g}}$, the amplitude $\mathbf{A}$, and the duration T) which serve as meta-parameters.

CrKR explores locally similar to PoWER. It can include an initial policy in the form of a prior or initial samples. The policy is updated after each rollout. Instead of fixing a parametrized policy, the kernel hyper-parameters need to be determined, which results in a more flexible policy representation as demonstrated in comparisons with finite difference gradients and episodic reward-weighted regression. The hyper-parameters can be estimated from initial demonstrations. The update equation also includes a matrix inversion, which theoretically could be replaced by a sparse approximation. However, in the proposed setting the sample size was sufficiently small to allow full matrix inversion.

7.1.2 Applications

The applications focus mainly on dynamical systems motor primitives as policy representations but we also demonstrated the approaches using various different parametrization. All approaches have been extensively evaluated on benchmark tasks as well as with simulated and real robots, namely a Barrett WAM, a BioRob, the JST-ICORP/SARCOS CBi and a Kuka KR 6.

We successfully learned the single movements of Tetherball Target Hitting, Casting, Underactuated Swing-Up, and Ball-in-a-Cup using imitation learning and subsequent reinforcement learning. Compared to alternative approaches the proposed algorithm consistently converged to a better solution in less rollouts. In the Ball-in-a-Cup task the number of optimized parameters would result in an unrealistic amount of rollouts for gradient based policy search approaches.

We learned to generalize the motor skills of dart throwing and table tennis strikes. In contrast to previous work on generalizing dynamical systems motor primitives, we employ reinforcement learning to discover a non-intuitive mapping between states and actions. We compared the proposed non-parametric approach to parametric ones and showed that it is hard to find a good parametrization in this setting.

We have demonstrated initial steps towards hierarchical learning with a ball target throwing task. In this task we learn how to hit targets while keeping in mind a higher level strategy. Finally, we employed a ball-bouncing task to explore first ideas towards learning to perform a task based on the concurrent execution of several motor primitives.

7.2   Open Problems

In this book, we have contributed to the state of art of autonomous acquisition of motor skill by robots by developing approaches for learning motor primitives via reinforcement learning in robotics. In the following we briefly discuss possible extensions of the presented work and future lines of research.

Learning Motor Skills

The presented learning approaches have been applied to episodic tasks that correspond to sports and games. In the following we present several ideas to render them applicable to a wider variety of tasks and how to overcome some of the current limitations.

Rhythmic Motor Tasks. This book focused on discrete motor tasks in an episodic setting. However, there are also many interesting rhythmic motor task that could be learned, e.g., the ball paddling movement, briefly discussed in Section 4.4.2, could be optimized to consume less energy. Several periods of such rhythmic tasks could be grouped into rollouts and the presented algorithms would be directly applicable. However, it is not obvious how such tasks could be learned online without frequent restarts. Deciding when complete restarts are required is an interesting challenge. Rhythmic tasks often require a start-up phase until transitioning into a stable rhythmic behavior. Similarly, even if a stable behavior with a high reward is achieved after online learning and exploration, it is not obvious to ensure that this state is directly reachable.

Compliance. The presented policy parametrizations are all based on positions, velocities and accelerations in joint or task space. Force controlled robots additionally allow to adapt their compliance and stiffness during the movement. When humans learn a new motor skill, they are often fairly tense initially but soon start to figure out during which parts of the movement they can relax their muscles. Decreasing the stiffness of the robot renders direct interactions with humans and its environment safer and has the additional benefit of being more energy efficient. Hence, the robot needs to remain compliant unless the specific part of the movement requires higher forces or precise positioning. Extending the dynamical systems motor primitives to include global or directional stiffness would render the presented algorithms applicable in this setting. A related approach has already been discussed by Kalakrishnan et al (2011) for this problem.

Sparse Approximations. We employed CrKR (Chapter 5) to generalize motor skills to new situations. In the current implementation we rely on the fact that the sample size is typically small enough to compute the updates in real-time. In Chapter 6 we employed a related approach to learn prioritized control, which is straightforward to kernelize. However, only a few seconds of training data could be used for real-time computations due to the high sampling rate. The matrix inversion is the main bottleneck, but also calculating the kernel between the current state and all stored samples can pose problems. As an alternative a local approximation or a sparse approximation could be considered (Nguyen-Tuong and Peters, 2011). For the local models the state-space is divided into smaller local ones, hence reducing the number of training points. In a sparse approximation only the most informative points are considered. The most challenging question for these approaches remains how to construct the local models or to decide which points to discard, include, or merge.

Continuous Adaptation. In Chapter 5, we evaluated tasks that only required a one-time adaptation to the situation. Especially for the table tennis task, continuously adapting the hitting movement to the ball's movement would render the task more reliable. Ball spin is hard to determine from the vision data before the ball bounces. However, due to acceleration limits of the robot, the hitting movement needs to be initiated before the bounce. Conceptually CrKR can handle this setting as well, however considering the whole state-space will result in significantly more samples, rendering real-time computations more challenging (see above), and also make generalizations more difficult due to the curse of dimensionality. Experiments need to be done in order to evaluate the feasibility and to determine whether additional components such as dimensionality reduction or models would be beneficial.

Manipulation. Compared to the presented applications (related to games and sports), grasping and manipulation often require less dynamic movements but in contrast have a higher dimensional state space due to the more complicated interactions between the robot and the objects. Grasping and manipulation are essential building blocks for many household tasks and have to be able to generalize to novel objects and tools. For example, opening a door needs to take into account a wide variety of handles and would benefit from adaptive compliance levels (as discussed above) for approaching the handle, pushing the handle, and opening the door.

Using Models. In Chapter 2, we identified the use of models as one of the promising approaches to make reinforcement learning for robotics tractable. We mainly employed a model of the robot and environment for debugging purposes, before testing the algorithms on the real robot. The approaches indirectly employ a model of the robot by using, e.g., the inverse dynamics controller of the robot. Pre-training in simulation can be helpful if the models are sufficiently accurate. An interesting extension of the proposed approaches would be to employ mental rehearsal with learned models in between real robot rollouts.

Hierarchical Reinforcement Learning

The hierarchical reinforcement learning approach presented in Chapter 5 is very specific to the problem setting. A more general hierarchical reinforcement learning approach would need the ability to choose between different motor primitives, the ability to discover new options, i.e., decide when it is better to generate a new one than to generalize an old one, and the ability to learn the shape parameters jointly with the meta-parameters. In a table tennis scenario we could have a library of already learned motor primitives. The hierarchical reinforcement learning approach would need to pick the most appropriate primitive (a forehand, a backhand, a smash, etc.) from this library according to the incoming ball, the current state of the robot, as well as strategic considerations. The template primitives will have to be generalized according to the current situation, either based on a single primitive or by combining several primitives. An approach based on the mixture of motor primitives has been recently proposed by Muelling et al (2010). Especially at the beginning of the learning process it will often be advantageous to learn and add a new primitive to the library instead of generalizing an old one.

Sequencing Motor Primitives. Another interesting direction for research is learning to optimally transition between motor primitives, e.g., to change from a forehand to a backhand. The transition needs to take into account external constraints such as self-collision and decide whether there is sufficient time for a successful transition. This problem is highly related to motion blending (Kovar, 2004). It would be very interesting to see how well the approaches developed by the computer graphics community transfer to robotics.

Superposition of Motor Primitives. In our daily lives we often perform several motor skills simultaneously like balancing a tray while walking or sidestepping and keeping balance while hitting a table tennis ball. In order to perform multiple motor skills simultaneously a system could for example overly, combine and prioritize motor primitives. First ideas have been explored in Chapter 6. Similar to inverse reinforcement learning it would be interesting to try to recover a dominance structure of primitives from human demonstrations. Alternatively, reinforcement learning could be employed to determine the best ordering for a given reward.

Learning Layers Jointly. For the ball-bouncing task (Chapter 6), we assumed that both the motor primitives and the strategy layer are fixed. In this task, better performance could be achieved if not only the prioritized control would be adapted but the primitives and the strategy as well. For example, the primitives could be adapted to compensate for shortcomings of the learned or the strategy could be adapted. A straightforward idea would be to run a reinforcement learning algorithm after the prioritized control has been learned. PoWER could be employed to learn the parameters of the underlying primitives. The approach presented in Chapter 5 is a first attempt to tackle this kind of hierarchical problem.

Multiple Strategies. Learning single motor skills with PoWER (Chapter 4) relied on a single initial demonstration. Combining several demonstrations can potentially provide a better understanding of the important features of the movement. However, using a weighted average of multiple demonstrations of different strategies is unlikely to succeed. Keeping multiple strategies in mind and figuring out which parts of the movement can be learned jointly and which parts have to be learned separately might lead to faster and more robust learning. Daniel et al (2012) employ a hierarchical reinforcement learning framework to learn different strategies for a thetherball task jointly.

References

Abbeel, P., Ng, A.Y.: Apprenticeship learning via inverse reinforcement learning. In: Proceedings of the 21st International Conference on Machine Learning (ICML), p. 1 (2004)

Abbeel, P., Quigley, M., Ng, A.Y.: Using inaccurate models in reinforcement learning. In: Proceedings of the International Conference on Machine Learning, ICML (2006)

Abbeel, P., Coates, A., Quigley, M., Ng, A.Y.: An application of reinforcement learning to aerobatic helicopter flight. In: Advances in Neural Information Processing Systems, NIPS (2007)

An, C.H., Atkeson, C.G., Hollerbach, J.M.: Model-based control of a robot manipulator. MIT Press, Cambridge (1988)

Andersson, R.: A robot ping-pong player: experiment in real-time intelligent control. MIT Press (1988)

Andrieu, C., de Freitas, N., Doucet, A., Jordan, M.I.: An introduction to MCMC for machine learning. Machine Learning 50(1), 5–43 (2003)

Argall, B.D., Browning, B., Veloso, M.: Learning robot motion control with demonstration and advice-operators. In: Proceedings of the IEEE/RSJ International Conference on Intelligent Robots and Systems, IROS (2008)

Argall, B.D., Chernova, S., Veloso, M., Browning, B.: A survey of robot learning from demonstration. Robotics and Autonomous Systems 57, 469–483 (2009)

Arimoto, S., Kawamura, S., Miyazaki, F.: Bettering operation of robots by learning. Journal of Robotic Systems 1(2), 123–140 (1984)

Asada, M., Noda, S., Tawaratsumida, S., Hosoda, K.: Purposive behavior acquisition for a real robot by vision-based reinforcement learning. Machine Learning 23(2-3), 279–303 (1996)

Atkeson, C.G.: Using local trajectory optimizers to speed up global optimization in dynamic programming. In: Cowan, J.D., Tesauro, G., Alspector, J. (eds.) Advances in Neural Information Processing Systems 6 (NIPS), Denver, CO, USA, pp. 503–521 (1994)

Atkeson, C.G.: Nonparametric model-based reinforcement learning. In: Advances in Neural Information Processing Systems, NIPS (1998)

Atkeson, C.G., Schaal, S.: Robot learning from demonstration. In: Proceedings of the International Conference on Machine Learning, ICML (1997)

Atkeson, C.G., Moore, A., Stefan, S.: Locally weighted learning for control. AI Review 11, 75–113 (1997)

Attias, H.: Planning by probabilistic inference. In: Proceedings of the 9th International Workshop on Artificial Intelligence and Statistics, AISTATS (2003)

Bagnell, J.A.: Learning decisions: Robustness, uncertainty, and approximation. PhD thesis, Robotics Institute, Carnegie Mellon University, Pittsburgh, PA (2004)

Bagnell, J.A., Schneider, J.C.: Autonomous helicopter control using reinforcement learning policy search methods. In: Proceedings of the IEEE International Conference on Robotics and Automation, ICRA (2001)

Bagnell, J.A., Schneider, J.C.: Covariant policy search. In: Proceedings of the International Joint Conference on Artificial Intelligence (IJCAI), pp. 1019–1024 (2003)

Bagnell, J.A., Ng, A.Y., Kakade, S., Schneider, J.C.: Policy search by dynamic programming. In: Advances in Neural Information Processing Systems 16, NIPS (2003)

Baird, L.C., Klopf, H.: Reinforcement learning with high-dimensional continuous actions. Tech. Rep. WL-TR-93-1147, Wright Laboratory, Wright-Patterson Air Force Base, OH 45433-7301 (1993)

Bakker, B., Zhumatiy, V., Gruener, G., Schmidhuber, J.: A robot that reinforcement-learns to identify and memorize important previous observations. In: Proceedings of the IEEE/RSJ International Conference on Intelligent Robots and Systems, IROS (2003)

Bakker, B., Zhumatiy, V., Gruener, G., Schmidhuber, J.: Quasi-online reinforcement learning for robots. In: Proceedings of the IEEE International Conference on Robotics and Automation, ICRA (2006)

Barto, A.G., Mahadevan, S.: Recent advances in hierarchical reinforcement learning. Discrete Event Dynamic Systems 13(4), 341–379 (2003)

Bays, P.M., Wolpert, D.M.: Computational principles of sensorimotor control that minimise uncertainty and variability. Journal of Physiology 578, 387–396 (2007)

Bellman, R.E.: Dynamic Programming. Princeton University Press, Princeton (1957)

Bellman, R.E.: Introduction to the Mathematical Theory of Control Processes, vol. 40-I. Academic Press, New York (1967)

Bellman, R.E.: Introduction to the Mathematical Theory of Control Processes, vol. 40-II. Academic Press, New York (1971)

Benbrahim, H., Franklin, J.A.: Biped dynamic walking using reinforcement learning. Robotics and Autonomous Systems 22(3-4), 283–302 (1997)

Benbrahim, H., Doleac, J.S., Franklin, J.A., Selfridge, O.: Real-time learning: A ball on a beam. In: Proceedings of the International Joint Conference on Neural Networks, IJCNN (1992)

Bentivegna, D.C.: Learning from observation using primitives. PhD thesis, Georgia Institute of Technology (2004)

Bentivegna, D.C., Atkeson, C.G., Cheng, G.: Learning from observation and practice using behavioral primitives: Marble maze (2004a)

Bentivegna, D.C., Ude, A., Atkeson, C.G., Cheng, G.: Learning to act from observation and practice. International Journal of Humanoid Robotics 1(4), 585–611 (2004)

Berg, J., Miller, S., Duckworth, D., Hu, H., Wan, A., Fu, X.Y., Goldberg, K., Abbeel, P.: Superhuman performance of surgical tasks by robots using iterative learning from human-guided demonstrations. In: Proceedings of the International Conference on Robotics and Automation, ICRA (2010)

Bertsekas, D.P.: Dynamic Programming and Optimal Control. Athena Scientific (1995)

Betts, J.T.: Practical methods for optimal control using nonlinear programming. In: Advances in Design and Control, Philadelphia, PA. Society for Industrial and Applied Mathematics (SIAM), vol. 3 (2001)

Binder, J., Koller, D., Russell, S.J., Kanazawa, K.: Adaptive probabilistic networks with hidden variables. Machine Learning 29(2-3), 213–244 (1997)

Birdwell, N., Livingston, S.: Reinforcement learning in sensor-guided AIBO robots. Tech. rep., University of Tennesse, Knoxville, advised by Dr. Itamar Elhanany (2007)

Bishop, C.M.: Pattern Recognition and Machine Learning. Information Science and Statistics. Springer (2006)

Bitzer, S., Howard, M., Vijayakumar, S.: Using dimensionality reduction to exploit constraints in reinforcement learning. In: Proceedings of the IEEE/RSJ International Conference on Intelligent Robots and Systems, IROS (2010)

Bootsma, R., van Wieringen, P.: Timing an attacking forehand drive in table tennis. Journal of Experimental Psychology: Human Perception and Performance 16, 21–29 (1990)

BotJunkie, Botjunkie interview: Nancy Dussault Smith on iRobot's roomba (2012), http://www.botjunkie.com/2010/05/17/botjunkie-interview-nancy-dussault-smith-on-irobots-roomba/

Boularias, A., Kober, J., Peters, J.: Relative entropy inverse reinforcement learning. Journal of Machine Learning Research - Proceedings Track 15, 182–189 (2011)

Boyan, J.A., Moore, A.W.: Generalization in reinforcement learning: Safely approximating the value function. In: Advances in Neural Information Processing Systems 7 (NIPS), pp. 369–376 (1994)

Brafman, R.I., Tennenholtz, M.: R-max - a general polynomial time algorithm for near-optimal reinforcement learning. Journal of Machine Learning Research 3, 213–231 (2002)

Buchli, J., Stulp, F., Theodorou, E., Schaal, S.: Learning variable impedance control. International Journal of Robotic Research 30(7), 820–833 (2011)

Buehler, M., Koditschek, D.E., Kindlmann, P.J.: Planning and control of robotic juggling and catching tasks. International Journal of Robotic Research 13(2), 101–118 (1994)

Bukkems, B.H.M., Kostic, D., de Jager, A.G., Steinbuch, M.: Learning-based identification and iterative learning control of direct-drive robots. IEEE Transactions on Control Systems Technology 13(4), 537–549 (2005)

Buşoniu, L., Babuška, R., De Schutter, B., Ernst, D.: Reinforcement Learning and Dynamic Programming Using Function Approximators. CRC Press, Boca Raton (2010)

Caruana, R.: Multitask learning. Machine Learning 28, 41–75 (1997)

Chan, H.W.K., King, I., Lui, J.: Performance analysis of a new updating rule for TD(λ) learning in feedforward networks for position evaluation in go game. In: Proceedings of the IEEE International Conference on Neural Networks, vol. 3, pp. 1716–1720 (1996)

Chaumette, F., Marchand, E.: A redundancy-based iterative approach for avoiding joint limits: application to visual servoing. IEEE Transactions on Robotics and Automation 17(5), 719–730 (2001)

Cheng, G., Hyon, S., Morimoto, J., Ude, A., Hale, J.G., Colvin, G., Scroggin, W., Jacobsen, S.C.: CB: A humanoid research platform for exploring neuroscience. Journal of Advance Robotics 21(10), 1097–1114 (2007)

Chiappa, S., Kober, J., Peters, J.: Using bayesian dynamical systems for motion template libraries. In: Advances in Neural Information Processing Systems 21 (NIPS), pp. 297–304 (2009)

Coates, A., Abbeel, P., Ng, A.Y.: Apprenticeship learning for helicopter control. Communications of the ACM 52(7), 97–105 (2009)

Cocora, A., Kersting, K., Plagemann, C., Burgard, W., Raedt, L.D.: Learning relational navigation policies. In: Proceedings of the IEEE/RSJ International Conference on Intelligent Robots and Systems, IROS (2006)

Conn, K., Peters II, R.A.: Reinforcement learning with a supervisor for a mobile robot in a real-world environment. In: Proceedings of the IEEE International Symposium on Computational Intelligence in Robotics and Automation, CIRA (2007)

Crammer, K., Dekel, O., Keshet, J., Shalev-Shwartz, S., Singer, Y.: Online passive-aggressive algorithms. Journal of Machine Learning Research 7, 551–585 (2006)

Daniel, C., Neumann, G., Peters, J.: Hierarchical relative entropy policy search. In: Proceedings of the International Conference on Artificial Intelligence and Statistics, AISTATS (2012)

DARPA, Autonomous robot manipulation, ARM (2010a),
`http://www.darpa.mil/ipto/programs/arm/arm.asp`

DARPA, Learning applied to ground robotics, LAGR (2010b),
`http://www.darpa.mil/ipto/programs/lagr/lagr.asp`

DARPA, Learning locomotion (L2) (2010c),
`http://www.darpa.mil/ipto/programs/ll/ll.asp`

Daumé III, H., Langford, J., Marcu, D.: Search-based structured prediction. Machine Learning Journal (MLJ) 75, 297–325 (2009)

Dayan, P., Hinton, G.E.: Using expectation-maximization for reinforcement learning. Neural Computation 9(2), 271–278 (1997)

Deisenroth, M.P., Rasmussen, C.E.: PILCO: A model-based and data-efficient approach to policy search. In: Proceedings of the 28th International Conference on Machine Learning (ICML), pp. 465–472 (2011)

Deisenroth, M.P., Rasmussen, C.E., Fox, D.: Learning to control a low-cost manipulator using data-efficient reinforcement learning. In: Proceedings of the 9th International Conference on Robotics: Science & Systems, R:SS (2011)

Dempster, A.P., Laird, N.M., Rubin, D.B.: Maximum likelihood from incomplete data via the EM algorithm. Journal of the Royal Statistical Society, Series B (Methodological) 39, 1–38 (1977)

Donnart, J.Y., Meyer, J.A.: Learning reactive and planning rules in a motivationally autonomous animat. IEEE Transactions on Systems, Man, and Cybernetics, Part B: Cybernetics 26(3), 381–395 (1996)

Donoho, D.L.: High-dimensional data analysis: the curses and blessings of dimensionality. In: American Mathematical Society Conference Math Challenges of the 21st Century (2000)

Dorigo, M., Colombetti, M.: Robot shaping: Developing situated agents through learning. Tech. rep., International Computer Science Institute, Berkeley, CA (1993)

Doya, K.: Metalearning and neuromodulation. Neural Networks 15(4-6), 495–506 (2002)

Duan, Y., Liu, Q., Xu, X.: Application of reinforcement learning in robot soccer. Engineering Applications of Artificial Intelligence 20(7), 936–950 (2007)

Duan, Y., Cui, B., Yang, H.: Robot navigation based on fuzzy RL algorithm. In: Proceedings of the International Symposium on Neural Networks, ISNN (2008)

El-Fakdi, A., Carreras, M., Ridao, P.: Towards direct policy search reinforcement learning for robot control. In: Proceedings of the IEEE/RSJ International Conference on Intelligent Robots and Systems, IROS (2006)

Endo, G., Morimoto, J., Matsubara, T., Nakanishi, J., Cheng, G.: Learning CPG-based biped locomotion with a policy gradient method: Application to a humanoid robot. International Journal of Robotics Research 27(2), 213–228 (2008)

Engel, Y., Mannor, S., Meir, R.: Reinforcement learning with gaussian processes. In: Proceedings of the International Conference on Machine Learning (ICML), pp. 201–208 (2005)

Erden, M.S., Leblebicioğlu, K.: Free gait generation with reinforcement learning for a six-legged robot. Robotics and Autonomous Systems (RAS) 56(3), 199–212 (2008)

Fagg, A.H., Lotspeich, D.L., Hoff, J., Bekey, G.A.: Rapid reinforcement learning for reactive control policy design for autonomous robots. In: Proceedings of Artificial Life in Robotics (1998)

Fantoni, I., Lozano, R.: Non-Linear Control for Underactuated Mechanical Systems. Springer-Verlag New York, Inc., Secaucus (2001)

Farley, B.G., Clark, W.A.: Simulation of self-organizing systems by digital computer. IRE Transactions on Information Theory 4, 76–84 (1954)

Fässler, H., Beyer, H.A., Wen, J.T.: A robot ping pong player: optimized mechanics, high performance 3d vision, and intelligent sensor control. Robotersysteme 6(3), 161–170 (1990)

Fidelman, P., Stone, P.: Learning ball acquisition on a physical robot. In: International Symposium on Robotics and Automation, ISRA (2004)

Findeisen, W., Bailey, F.N., Brdeys, M., Malinowski, K., Tatjewski, P., Wozniak, J.: Control and coordination in hierarchical systems. International series on applied systems analysis. J. Wiley, Chichester (1980)

Freeman, C., Lewin, P., Rogers, E., Ratcliffe, J.: Iterative learning control applied to a gantry robot and conveyor system. Transactions of the Institute of Measurement and Control 32(3), 251–264 (2010)

Gams, A., Ude, A.: Generalization of example movements with dynamic systems. In: Proceedings of the IEEE/RSJ International Conference on Humanoid Robots, HUMANOIDS (2009)

Gaskett, C., Fletcher, L., Zelinsky, A.: Reinforcement learning for a vision based mobile robot. In: Proceedings of the IEEE/RSJ International Conference on Intelligent Robots and Systems, IROS (2000)

Geng, T., Porr, B., Wörgötter, F.: Fast biped walking with a reflexive controller and real-time policy searching. In: Advances in Neural Information Processing Systems, NIPS (2006)

Glynn, P.: Likelihood ratio gradient estimation: An overview. In: Proceedings of the Winter Simulation Conference, WSC (1987)

Goldberg, D.E.: Genetic algorithms. Addision Wesley (1989)

Gordon, G.J.: Approximate solutions to markov decision processes. PhD thesis, School of Computer Science, Carnegie Mellon University (1999)

Gräve, K., Stückler, J., Behnke, S.: Learning motion skills from expert demonstrations and own experience using Gaussian process regression. In: Proceedings of the Joint International Symposium on Robotics (ISR) and German Conference on Robotics, ROBOTIK (2010)

Greensmith, E., Bartlett, P.L., Baxter, J.: Variance reduction techniques for gradient estimates in reinforcement learning. Journal of Machine Learning Research 5, 1471–1530 (2004)

Grimes, D.B., Rao, R.P.N.: Learning nonparametric policies by imitation. In: Proceedings of the IEEE/RSJ International Conference on Intelligent Robots and System (IROS), pp. 2022–2028 (2008)

Guenter, F., Hersch, M., Calinon, S., Billard, A.: Reinforcement learning for imitating constrained reaching movements. Advanced Robotics, Special Issue on Imitative Robots 21(13), 1521–1544 (2007)

Gullapalli, V., Franklin, J.A., Benbrahim, H.: Aquiring robot skills via reinforcement learning. IEEE Control Systems Journal, Special Issue on Robotics: Capturing Natural Motion 14(1), 13–24 (1994)

Hafner, R., Riedmiller, M.: Reinforcement learning on a omnidirectional mobile robot. In: Proceedings of the IEEE/RSJ International Conference on Intelligent Robots and Systems, IROS (2003)

Hafner, R., Riedmiller, M.: Neural reinforcement learning controllers for a real robot application. In: Proceedings of the IEEE International Conference on Robotics and Automation, ICRA (2007)

Hailu, G., Sommer, G.: Integrating symbolic knowledge in reinforcement learning. In: Proceedings of the IEEE International Conference on Systems, Man and Cybernetics, SMC (1998)

Hart, S., Grupen, R.: Learning generalizable control programs. IEEE Transactions on Autonomous Mental Development 3(3), 216–231 (2011)

Hester, T., Quinlan, M., Stone, P.: Generalized model learning for reinforcement learning on a humanoid robot. In: Proceedings of the IEEE International Conference on Robotics and Automation, ICRA (2010)

Hester, T., Quinlan, M., Stone, P.: RTMBA: A real-time model-based reinforcement learning architecture for robot control. In: IEEE International Conference on Robotics and Automation, ICRA (2012)

Hoffman, M., Doucet, A., de Freitas, N., Jasra, A.: Bayesian policy learning with trans-dimensional MCMC. In: Advances in Neural Information Processing Systems 20, NIPS (2007)

Huang, X., Weng, J.: Novelty and reinforcement learning in the value system of developmental robots. In: Proceedings of the 2nd International Workshop on Epigenetic Robotics: Modeling Cognitive Development in Robotic Systems (2002)

Hubbard, A., Seng, C.: Visual movements of batters. Research Quaterly 25, 42–57 (1954)

Huber, M., Grupen, R.: Learning robot control – using control policies as abstract actions. In: Proceedings of the NIPS 1998 Workshop: Abstraction and Hierarchy in Reinforcement Learning (1998)

Huber, M., Grupen, R.A.: A feedback control structure for on-line learning tasks. Robotics and Autonomous Systems 22(3-4), 303–315 (1997)

Ijspeert, A.J., Nakanishi, J., Schaal, S.: Learning attractor landscapes for learning motor primitives. In: Advances in Neural Information Processing Systems 15 (NIPS), pp. 1523–1530 (2002a)

Ijspeert, A.J., Nakanishi, J., Schaal, S.: Movement imitation with nonlinear dynamical systems in humanoid robots. In: Proceedings of the IEEE International Conference on Robotics and Automation (ICRA), Washington, DC, pp. 1398–1403 (2002b)

Ilg, W., Albiez, J., Jedele, H., Berns, K., Dillmann, R.: Adaptive periodic movement control for the four legged walking machine BISAM. In: Proceedings of the IEEE International Conference on Robotics and Automation, ICRA (1999)

Jaakkola, T., Jordan, M.I., Singh, S.P.: Convergence of stochastic iterative dynamic programming algorithms. In: Advances in Neural Information Processing Systems (NIPS), vol. 6, pp. 703–710 (1993)

Jacobson, D.H., Mayne, D.Q.: Differential Dynamic Programming. Elsevier (1970)

Jakobi, N., Husbands, P., Harvey, I.: Noise and the reality gap: The use of simulation in evolutionary robotics. In: Proceedings of the 3rd European Conference on Artificial Life, pp. 704–720 (1995)

Jetchev, N., Toussaint, M.: Trajectory prediction: learning to map situations to robot trajectories. In: Proceedings of the International Conference on Machine Learning (ICML), p. 57 (2009)

Kaelbling, L.P.: Learning in embedded systems. PhD thesis, Stanford University, Stanford, California (1990)

Kaelbling, L.P., Littman, M.L., Moore, A.W.: Reinforcement learning: A survey. Journal of Artificial Intelligence Research 4, 237–285 (1996)

Kakade, S.: On the sample complexity of reinforcement learning. PhD thesis, Gatsby Computational Neuroscience Unit. University College London (2003)

Kakade, S., Langford, J.: Approximately optimal approximate reinforcement learning. In: Proceedings of the 19th International Conference on Machine Learning (ICML), pp. 267–274 (2002)

Kalakrishnan, M., Righetti, L., Pastor, P., Schaal, S.: Learning force control policies for compliant manipulation. In: Proceedings of the IEEE/RSJ International Conference on Intelligent Robots and Systems (IROS), pp. 4639–4644 (2011)

Kalman, R.: A new approach to linear filtering and prediction problems. Transactions of the ASME–Journal of Basic Engineering 82(Series D), 35–45 (1960)

Kalman, R.E.: When is a linear control system optimal? Journal of Basic Engineering 86(1), 51–60 (1964)

Kalmár, Z., Szepesvári, C., Lőrincz, A.: Modular reinforcement learning: An application to a real robot task. In: Birk, A., Demiris, J. (eds.) EWLR 1997. LNCS (LNAI), vol. 1545, pp. 29–45. Springer, Heidelberg (1998)

Kappen, H.J.: Path integrals and symmetry breaking for optimal control theory. Journal of Statistical Mechanics: Theory and Experiment 11, P11,011 (2005)

Katz, D., Pyuro, Y., Brock, O.: Learning to manipulate articulated objects in unstructured environments using a grounded relational representation. In: Proceedings of Robotics: Science and Systems Conference, R:SS (2008)

Kawato, M.: Feedback-error-learning neural network for supervised motor learning. Advanced Neural Computers 6(3), 365–372 (1990)

Kearns, M., Singh, S.P.: Near-optimal reinforcement learning in polynomial time. Machine Learning 49(2-3), 209–232 (2002)

Keeney, R., Raiffa, H.: Decisions with multiple objectives: Preferences and value tradeoffs. J. Wiley, New York (1976)

Khatib, O.: Real-time obstacle avoidance for manipulators and mobile robots. International Journal of Robotics Research 5(1), 90–98 (1986)

Kimura, H., Yamashita, T., Kobayashi, S.: Reinforcement learning of walking behavior for a four-legged robot. In: Proceedings of the IEEE Conference on Decision and Control, CDC (2001)

Kirchner, F.: Q-learning of complex behaviours on a six-legged walking machine. In: Proceedings of the EUROMICRO Workshop on Advanced Mobile Robots (1997)

Kirk, D.E.: Optimal control theory. Prentice-Hall, Englewood Cliffs (1970)

Ko, J., Klein, D.J., Fox, D., Hähnel, D.: Gaussian processes and reinforcement learning for identification and control of an autonomous blimp. In: Proceedings of the IEEE International Conference on Robotics and Automation, ICRA (2007)

Kober, J., Peters, J.: Policy search for motor primitives in robotics. In: Advances in Neural Information Processing Systems 21 (NIPS), pp. 849–856 (2008)

Kober, J., Peters, J.: Learning motor primitives for robotics. In: Proceedings of the IEEE International Conference on Robotics and Automation (ICRA), pp. 2112–2118 (2009)

Kober, J., Peters, J.: Imitation and reinforcement learning - practical algorithms for motor primitive learning in robotics. IEEE Robotics and Automation Magazine 17(2), 55–62 (2010)

Kober, J., Peters, J.: Learning elementary movements jointly with a higher level task. In: Proceedings of the IEEE/RSJ International Conference on Intelligent Robots and Systems (IROS), pp. 338–343 (2011a)

Kober, J., Peters, J.: Policy search for motor primitives in robotics. Machine Learning 84(1-2), 171–203 (2011b)

Kober, J., Peters, J.: Reinforcement Learning in Robotics: A Survey. In: Wiering, M., van Otterlo, M. (eds.) Reinforcement Learning. ALO, vol. 12, pp. 579–610. Springer, Heidelberg (2012)

Kober, J., Peters, J.: Learning Prioritized Control of Motor Primitives. arXiv:1209.0488 (cs.RO) (2012)

Kober, J., Mohler, B., Peters, J.: Learning perceptual coupling for motor primitives. In: Proceedings of the IEEE/RSJ International Conference on Intelligent Robots and Systems (IROS), pp. 834–839 (2008)

Kober, J., Muelling, K., Kroemer, O., Lampert, C.H., Schölkopf, B., Peters, J.: Movement templates for learning of hitting and batting. In: Proceedings of the IEEE International Conference on Robotics and Automation (ICRA), pp. 853–858 (2010a)

Kober, J., Oztop, E., Peters, J.: Reinforcement learning to adjust robot movements to new situations. In: Proceedings of Robotics: Science and Systems Conference, R:SS (2010b)

Kober, J., Oztop, E., Peters, J.: Reinforcement learning to adjust robot movements to new situations. In: Proceedings of the 22nd International Joint Conference on Artificial Intelligence (IJCAI), Best Paper Track, pp. 2650–2655 (2011)

Kober, J., Wilhelm, A., Oztop, E., Peters, J.: Reinforcement learning to adjust parametrized motor primitives to new situations. Autonomous Robots 33(4), 361–379 (2012)

Kober, J., Bagnell, J.A., Peters, J.: Reinforcement learning in robotics: A survey. International Journal of Robotics Research 32(11), 1236–1272 (2013)

Kohl, N., Stone, P.: Policy gradient reinforcement learning for fast quadrupedal locomotion. In: Proceedings of the IEEE International Conference on Robotics and Automation, ICRA (2004)

Kollar, T., Roy, N.: Trajectory optimization using reinforcement learning for map exploration. International Journal of Robotics Research 27(2), 175–197 (2008)

Kolter, J.Z., Ng, A.Y.: Policy search via the signed derivative. In: Proceedings of Robotics: Science and Systems Conference, R:SS (2009a)

Kolter, J.Z., Ng, A.Y.: Regularization and feature selection in least-squares temporal difference learning. In: International Conference on Machine Learning, ICML (2009b)

Kolter, J.Z., Abbeel, P., Ng, A.Y.: Hierarchical apprenticeship learning with application to quadruped locomotion. In: Advances in Neural Information Processing Systems, NIPS (2007)

Kolter, J.Z., Coates, A., Ng, A.Y., Gu, Y., DuHadway, C.: Space-indexed dynamic programming: learning to follow trajectories. In: Proceedings of the International Conference on Machine Learning, ICML (2008)

Kolter, J.Z., Plagemann, C., Jackson, D.T., Ng, A.Y., Thrun, S.: A probabilistic approach to mixed open-loop and closed-loop control, with application to extreme autonomous driving. In: Proceedings of the IEEE International Conference on Robotics and Automation, ICRA (2010)

Konidaris, G.D., Kuindersma, S., Grupen, R., Barto, A.G.: Autonomous skill acquisition on a mobile manipulator. In: AAAI Conference on Artificial Intelligence (2011a)

Konidaris, G.D., Osentoski, S., Thomas, P.: Value function approximation in reinforcement learning using the Fourier basis. In: AAAI Conference on Artificial Intelligence, AAAI (2011b)

Konidaris, G.D., Kuindersma, S., Grupen, R., Barto, A.G.: Robot learning from demonstration by constructing skill trees. International Journal of Robotics Research 31(3), 360–375 (2012)

Kormushev, P., Calinon, S., Caldwell, D.G.: Robot motor skill coordination with em-based reinforcement learning. In: Proceedings of the IEEE/RSJ International Conference on Intelligent Robots and Systems, IROS (2010)

Kovar, L.: Automated methods for data-driven synthesis of realistic and controllable human motion. PhD thesis, University of Wisconsin at Madison (2004)

Kroemer, O., Detry, R., Piater, J., Peters, J.: Active learning using mean shift optimization for robot grasping. In: Proceedings of the IEEE/RSJ International Conference on Intelligent Robots and Systems, IROS (2009)

Kroemer, O., Detry, R., Piater, J., Peters, J.: Combining active learning and reactive control for robot grasping. Robotics and Autonomous Systems 58(9), 1105–1116 (2010)

Kronander, K., Khansari-Zadeh, M.S.M., Billard, A.: Learning to control planar hitting motions in a minigolf-like task. In: Proceedings of the IEEE/RSJ International Conference on Intelligent Robots and Systems (IROS), pp. 710–717 (2011)

Kuhn, H.W., Tucker, A.W.: Nonlinear programming. In: Proceedings of the Berkeley Symposium on Mathematical Statistics and Probability (1950)

Kuindersma, S., Grupen, R., Barto, A.G.: Learning dynamic arm motions for postural recovery. In: IEEE-RAS International Conference on Humanoid Robots, HUMANOIDS (2011)

Kulchenko, P.: Robot juggles two ping-pong balls (2011),
http://notebook.kulchenko.com/juggling/robot-juggles-two-ping-pong-balls

Kwee, I., Hutter, M., Schmidhuber, J.: Gradient-based reinforcement planning in policy-search methods. In: Proceedings of the 5th European Workshop on Reinforcement Learning (EWRL), vol. 27, pp. 27–29 (2001)

Kwok, C., Fox, D.: Reinforcement learning for sensing strategies. In: IEEE/RSJ International Conference on Intelligent Robots and Systems, IROS (2004)

Lampariello, R., Nguyen-Tuong, D., Castellini, C., Hirzinger, G., Peters, J.: Trajectory planning for optimal robot catching in real-time. In: Proceedings of the IEEE International Conference on Robotics and Automation (ICRA), pp. 3719–3726 (2011)

Langford, J., Zadrozny, B.: Relating reinforcement learning performance to classification performance. In: 22nd International Conference on Machine Learning, ICML (2005)

Latzke, T., Behnke, S., Bennewitz, M.: Imitative reinforcement learning for soccer playing robots. In: Lakemeyer, G., Sklar, E., Sorrenti, D.G., Takahashi, T. (eds.) RoboCup 2006. LNCS (LNAI), vol. 4434, pp. 47–58. Springer, Heidelberg (2007)

Laud, A.D.: Theory and application of reward shaping in reinforcement learning. PhD thesis, University of Illinois at Urbana-Champaign (2004)

Lawrence, G., Cowan, N., Russell, S.: Efficient gradient estimation for motor control learning. In: Proceedings of the International Conference on Uncertainty in Artificial Intelligence (UAI), pp. 354–361 (2003)

Lens, T., Kunz, J., Trommer, C., Karguth, A., von Stryk, O.: Biorob-arm: A quickly deployable and intrinsically safe, light-weight robot arm for service robotics applications. In: Proceedings of the 41st International Symposium on Robotics / 6th German Conference on Robotics, pp. 905–910 (2010)

Lewis, M.E., Puterman, M.L.: Bias Optimality. In: The Handbook of Markov Decision Processes: Methods and Applications, pp. 89–111. Kluwer (2001)

Lin, H.I., Lai, C.C.: Learning collision-free reaching skill from primitives. In: IEEE/RSJ International Conference on Intelligent Robots and Systems, IROS (2012)

Lizotte, D., Wang, T., Bowling, M., Schuurmans, D.: Automatic gait optimization with Gaussian process regression. In: Proceedings of the International Joint Conference on Artifical Intelligence, IJCAI (2007)

van der Maaten, L.J.P., Postma, E.O., van den Herik, H.J.: Dimensionality reduction: A comparative review. Tech. Rep. TiCC-TR 2009-005, Tilburg University (2009)

Mahadevan, S., Connell, J.: Automatic programming of behavior-based robots using reinforcement learning. Artificial Intelligence 55(2-3), 311–365 (1992)

Martín, H.J., de Lope, J., Maravall, D.: The kNN-TD reinforcement learning algorithm. In: Proceedings of the 3rd International Work-Conference on the Interplay Between Natural and Artificial Computation (IWINAC), pp. 305–314 (2009)

Martínez-Marín, T., Duckett, T.: Fast reinforcement learning for vision-guided mobile robots. In: Proceedings of the IEEE International Conference on Robotics and Automation, ICRA (2005)

Masters Games Ltd. The rules of darts (2010),
  `http://www.mastersgames.com/rules/darts-rules.htm`

Matarić, M.J.: Reward functions for accelerated learning. In: Proceedings of the International Conference on Machine Learning, ICML (1994)

Matarić, M.J.: Reinforcement learning in the multi-robot domain. Autonomous Robots 4, 73–83 (1997)

Matsushima, M., Hashimoto, T., Takeuchi, M., Miyazaki, F.: A learning approach to robotic table tennis. IEEE Transactions on Robotics 21(4), 767–771 (2005)

McGovern, A., Barto, A.G.: Automatic discovery of subgoals in reinforcement learning using diverse density. In: Proceedings of the International Conference on Machine Learning (ICML), pp. 361–368 (2001)

McGovern, A., Sutton, R.S., Fagg, A.H.: Roles of macro-actions in accelerating reinforcement learning. In: Proceedings of Grace Hopper Celebration of Women in Computing (1997)

McLachan, G.J., Krishnan, T.: The EM Algorithm and Extensions. Wiley Series in Probability and Statistics. John Wiley & Sons (1997)

Michels, J., Saxena, A., Ng, A.Y.: High speed obstacle avoidance using monocular vision and reinforcement learning. In: Proceedings of the International Conference on Machine Learning, ICML (2005)

Minsky, M.L.: Theory of neural-analog reinforcement systems and its application to the brain-model problem. PhD thesis, Princeton University (1954)

Mitsunaga, N., Smith, C., Kanda, T., Ishiguro, H., Hagita, N.: Robot behavior adaptation for human-robot interaction based on policy gradient reinforcement learning. In: Proceedings of the IEEE/RSJ International Conference on Intelligent Robots and Systems, IROS (2005)

Miyamoto, H., Schaal, S., Gandolfo, F., Gomi, H., Koike, Y., Osu, R., Nakano, E., Wada, Y., Kawato, M.: A Kendama learning robot based on bi-directional theory. Neural Networks 9(8), 1281–1302 (1996)

Moldovan, T.M., Abbeel, P.: Safe exploration in markov decision processes. In: 29th International Conference on Machine Learning (ICML), pp. 1711–1718 (2012)

Moore, A.W., Atkeson, C.G.: Prioritized sweeping: Reinforcement learning with less data and less time. Machine Learning 13(1), 103–130 (1993)

Morimoto, J., Doya, K.: Acquisition of stand-up behavior by a real robot using hierarchical reinforcement learning. Robotics and Autonomous Systems 36(1), 37–51 (2001)

Muelling, K., Peters, J.: A computational model of human table tennis for robot application. In: Proceedings of Autonome Mobile Systeme, AMS (2009)

Muelling, K., Kober, J., Peters, J.: Learning table tennis with a mixture of motor primitives. In: Proceedings of the 10th IEEE-RAS International Conference on Humanoid Robots (Humanoids), pp. 411–416 (2010)

Muelling, K., Kober, J., Peters, J.: A biomimetic approach to robot table tennis. Adaptive Behavior 9(5), 359–376 (2011)

Muelling, K., Kober, J., Kroemer, O., Peters, J.: Learning to select and generalize striking movements in robot table tennis. International Journal of Robotics Research (2012)

Müller, M., Lupashin, S., D'Andrea, R.: Quadrocopter ball juggling. In: Proceedings of the IEEE International Conference on Intelligent Robots and Systems (IROS), pp. 5113–5120 (2011)

Nakanishi, J., Morimoto, J., Endo, G., Cheng, G., Schaal, S., Kawato, M.: Learning from demonstration and adaptation of biped locomotion. Robotics and Autonomous Systems (RAS) 47(2-3), 79–91 (2004)

Nakanishi, J., Cory, R., Mistry, M., Peters, J., Schaal, S.: Operational space control: A theoretical and emprical comparison. International Journal of Robotics Research 27, 737–757 (2008)

Nemec, B., Tamošiūnaitė, M., Wörgötter, F., Ude, A.: Task adaptation through exploration and action sequencing. In: Proceedings of the IEEE-RAS International Conference on Humanoid Robots, HUMANOIDS (2009)

Nemec, B., Zorko, M., Zlajpah, L.: Learning of a ball-in-a-cup playing robot. In: Proceedings of the International Workshop on Robotics in Alpe-Adria-Danube Region, RAAD (2010)

Ng, A.Y., Jordan, M.: Pegasus: A policy search method for large mdps and pomdps. In: Proceedings of the International Conference on Uncertainty in Artificial Intelligence (UAI), pp. 406–415 (2000)

Ng, A.Y., Harada, D., Russell, S.J.: Policy invariance under reward transformations: Theory and application to reward shaping. In: Proceedings of the 16th International Conference on Machine Learning (ICML), pp. 278–287 (1999)

Ng, A.Y., Coates, A., Diel, M., Ganapathi, V., Schulte, J., Tse, B., Berger, E., Liang, E.: Autonomous inverted helicopter flight via reinforcement learning. In: Proceedings of the International Symposium on Experimental Robotics, ISER (2004a)

Ng, A.Y., Kim, H.J., Jordan, M.I., Sastry, S.: Inverted autonomous helicopter flight via reinforcement learning. In: Proceedings of the International Symposium on Experimental Robotics (ISER). MIT Press (2004b)

Nguyen-Tuong, D., Peters, J.: Incremental online sparsification for model learning in real-time robot control. Neurocomputing 74(11), 1859–1867 (2011)

Norrlöf, M.: An adaptive iterative learning control algorithm with experiments on an industrial robot. IEEE Transactions on Robotics 18(2), 245–251 (2002)

Oßwald, S., Hornung, A., Bennewitz, M.: Learning reliable and efficient navigation with a humanoid. In: Proceedings of the IEEE International Conference on Robotics and Automation, ICRA (2010)

Paletta, L., Fritz, G., Kintzler, F., Irran, J., Dorffner, G.: Perception and developmental learning of affordances in autonomous robots. In: Hertzberg, J., Beetz, M., Englert, R. (eds.) KI 2007. LNCS (LNAI), vol. 4667, pp. 235–250. Springer, Heidelberg (2007)

Park, D.H., Hoffmann, H., Pastor, P., Schaal, S.: Movement reproduction and obstacle avoidance with dynamic movement primitives and potential fields. In: Proceedings of the IEEE International Conference on Humanoid Robots (HUMANOIDS), pp. 91–98 (2008)

PASCAL2, Challenges (2010), http://pascallin2.ecs.soton.ac.uk/Challenges/

Pastor, P., Hoffmann, H., Asfour, T., Schaal, S.: Learning and generalization of motor skills by learning from demonstration. In: Proceedings of the IEEE International Conference on Robotics and Automation (ICRA), pp. 1293–1298 (2009)

Pastor, P., Kalakrishnan, M., Chitta, S., Theodorou, E., Schaal, S.: Skill learning and task outcome prediction for manipulation. In: Proceedings of the IEEE International Conference on Robotics and Automation, ICRA (2011)

Pendrith, M.: Reinforcement learning in situated agents: Some theoretical problems and practical solutions. In: Proceedings of the European Workshop on Learning Robots, EWRL (1999)

Peng, J., Williams, R.J.: Incremental multi-step q-learning. Machine Learning 22(1), 283–290 (1996)

Perkins, T.J., Barto, A.G.: Lyapunov design for safe reinforcement learning. Journal of Machine Learning Research 3, 803–832 (2002)

Peshkin, L.: Reinforcement learning by policy search. PhD thesis, Brown University, Providence, RI (2001)

Peters, J.: Machine learning of motor skills for robotics. PhD thesis, University of Southern California, Los Angeles, CA, 90089, USA (2007)

Peters, J., Schaal, S.: Policy gradient methods for robotics. In: Proceedings of the IEEE/RSJ International Conference on Intelligent Robots and Systems (IROS), pp. 2219–2225 (2006)

Peters, J., Schaal, S.: Reinforcement learning by reward-weighted regression for operational space control. In: Proceedings of the International Conference on Machine Learning, ICML (2007)

Peters, J., Schaal, S.: Learning to control in operational space. International Journal of Robotics Research 27(2), 197–212 (2008a)

Peters, J., Schaal, S.: Natural actor-critic. Neurocomputing 71(7-9), 1180–1190 (2008b)

Peters, J., Schaal, S.: Reinforcement learning of motor skills with policy gradients. Neural Networks 21(4), 682–697 (2008c)

Peters, J., Vijayakumar, S., Schaal, S.: Reinforcement learning for humanoid robotics. In: Proceedings of the IEEE-RAS International Conference on Humanoid Robots (HUMANOIDS), pp. 103–123 (2003)

Peters, J., Vijayakumar, S., Schaal, S.: Linear quadratic regulation as benchmark for policy gradient methods. Tech. rep., University of Southern California (2004)

Peters, J., Vijayakumar, S., Schaal, S.: Natural actor-critic. In: Proceedings of the European Conference on Machine Learning (ECML), pp. 280–291 (2005)

Peters, J., Muelling, K., Altun, Y.: Relative entropy policy search. In: Proceedings of the National Conference on Artificial Intelligence, AAAI (2010a)

Peters, J., Muelling, K., Kober, J., Nguyen-Tuong, D., Kroemer, O.: Towards motor skill learning for robotics. In: Proceedings of the International Symposium on Robotics Research, ISRR (2010b)

Piater, J.H., Jodogne, S., Detry, R., Kraft, D., Krüger, N., Kroemer, O., Peters, J.: Learning visual representations for perception-action systems. International Journal of Robotics Research 30(3), 294–307 (2011)

Platt, R., Grupen, R.A., Fagg, A.H.: Improving grasp skills using schema structured learning. In: Proceedings of the International Conference on Development and Learning (2006)

Pongas, D., Billard, A., Schaal, S.: Rapid synchronization and accurate phase-locking of rhythmic motor primitives. In: Proceedings of the IEEE/RSJ International Conference on Intelligent Robots and Systems (IROS), pp. 2911–2916 (2005)

Powell, W.B.: AI, OR and control theory: A rosetta stone for stochastic optimization. Tech. rep., Princeton University (2012)

Precup, D., Sutton, R.S., Singh, S.: Theoretical results on reinforcement learning with temporally abstract options. In: Proceedings of the European Confonference on Machine Learning, ECML (1998)

Puterman, M.L.: Markov Decision Processes: Discrete Stochastic Dynamic Programming. Wiley-Interscience (1994)

Ramanantsoa, M., Durey, A.: Towards a stroke construction model. International Journal of Table Tennis Science 2, 97–114 (1994)

Randløv, J., Alstrøm, P.: Learning to drive a bicycle using reinforcement learning and shaping. In: International Conference on Machine Learning (ICML), pp. 463–471 (1998)

Rasmussen, C.E., Williams, C.K.I.: Gaussian Processes for Machine Learning. Adaptive Computation And Machine Learning. MIT Press (2006)

Åström, K.J., Wittenmark, B.: Adaptive control. Addison-Wesley, Reading (1989)

Ratliff, N., Bradley, D., Bagnell, J.A., Chestnutt, J.: Boosting structured prediction for imitation learning. In: Advances in Neural Information Processing Systems, NIPS (2006a)

Ratliff, N., Bagnell, J.A., Srinivasa, S.: Imitation learning for locomotion and manipulation. In: Proceedings of the IEEE-RAS International Conference on Humanoid Robots, HUMANOIDS (2007)

Ratliff, N.D., Bagnell, J.A., Zinkevich, M.A.: Maximum margin planning. In: Proceedings of the 23rd International Conference on Machine Learning, ICML (2006b)

Reist, P., D'Andrea, R.: Bouncing an unconstrained ball in three dimensions with a blind juggling robot. In: Proceedings of the IEEE International Conference on Robotics and Automation (ICRA), pp. 2753–2760 (2009)

Riedmiller, M., Gabel, T., Hafner, R., Lange, S.: Reinforcement learning for robot soccer. Autonomous Robots 27(1), 55–73 (2009)

Rivlin, T.J.: An Introduction to the Approximation of Functions. Courier Dover Publications (1969)

Roberts, J.W., Moret, L., Zhang, J., Tedrake, R.: Motor learning at intermediate Reynolds number: experiments with policy gradient on the flapping flight of a rigid wing. In: Sigaud, O., Peters, J. (eds.) From Motor Learning to Interaction Learning in Robots. SCI, vol. 264, pp. 293–309. Springer, Heidelberg (2010)

Roberts, J.W., Manchester, I., Tedrake, R.: Feedback controller parameterizations for reinforcement learning. In: IEEE Symposium on Adaptive Dynamic Programming and Reinforcement Learning, ADPRL (2011)

Rosenstein, M.T., Barto, A.G.: Reinforcement learning with supervision by a stable controller. In: American Control Conference (2004)

Ross, S., Bagnell, J.A.: Agnostic system identification for model-based reinforcement learning. In: Proceedings of the International Conference on Machine Learning, ICML (2012)

Ross, S., Gordon, G., Bagnell, J.A.: A reduction of imitation learning and structured prediction to no-regret online learning. In: Proceedings of the 14th International Conference on Artifical Intelligence and Statistics, AISTATS (2011a)

Ross, S., Munoz, D., Hebert, M., AndrewBagnell, J.: Learning message-passing inference machines for structured prediction. In: Proceedings of the IEEE Conference on Computer Vision and Pattern Recognition, CVPR (2011b)

Rottmann, A., Plagemann, C., Hilgers, P., Burgard, W.: Autonomous blimp control using model-free reinforcement learning in a continuous state and action space. In: Proceedings of the IEEE/RSJ International Conference on Intelligent Robots and Systems, IROS (2007)

Rubinstein, R.Y., Kroese, D.P.: The Cross Entropy Method: A Unified Approach To Combinatorial Optimization, Monte-carlo Simulation (Information Science and Statistics). Springer-Verlag New York, Inc., Secaucus (2004)

Rückstieß, T., Felder, M., Schmidhuber, J.: State-dependent exploration for policy gradient methods. In: Proceedings of the European Conference on Machine Learning (ECML), pp. 234–249 (2008)

Russell, S.: Learning agents for uncertain environments (extended abstract). In: Proceedings of the 11th Annual Conference on Computational Learning Theory (COLT), pp. 101–103 (1998)

Rust, J.: Using randomization to break the curse of dimensionality. Econometrica 65(3), 487–516 (1997)

Sato, M., Nakamura, Y., Ishii, S.: Reinforcement learning for biped locomotion. In: Proceedings of the International Conference on Artificial Neural Networks, ICANN (2002)

Sato, S., Sakaguchi, T., Masutani, Y., Miyazaki, F.: Mastering of a task with interaction between a robot and its environment: "kendama" task. Transactions of the Japan Society of Mechanical Engineers C 59(558), 487–493 (1993)

Schaal, S.: Learning from demonstration. In: Advances in Neural Information Processing Systems, NIPS (1996)

Schaal, S.: Is imitation learning the route to humanoid robots? Trends in Cognitive Sciences 3(6), 233–242 (1999)

Schaal, S.: The SL simulation and real-time control software package. Tech. rep., University of Southern California (2009)

Schaal, S., Atkeson, C.G.: Robot juggling: An implementation of memory-based learning. Control Systems Magazine 14(1), 57–71 (1994)

Schaal, S., Sternad, D., Atkeson, C.G.: One-handed juggling: a dynamical approach to a rhythmic movement task. Journal of Motor Behavior 28(2), 165–183 (1996)

Schaal, S., Atkeson, C.G., Vijayakumar, S.: Scalable techniques from nonparameteric statistics for real-time robot learning. Applied Intelligence 17(1), 49–60 (2002)

Schaal, S., Peters, J., Nakanishi, J., Ijspeert, A.J.: Control, planning, learning, and imitation with dynamic movement primitives. In: Proceedings of the Workshop on Bilateral Paradigms on Humans and Humanoids, IEEE International Conference on Intelligent Robots and Systems, IROS (2003)

Schaal, S., Mohajerian, P., Ijspeert, A.J.: Dynamics systems vs. optimal control – A unifying view. Progress in Brain Research 165(1), 425–445 (2007)

Schmidt, R., Wrisberg, C.: Motor Learning and Performance, 2nd edn. Human Kinetics (2000)

Schneider, J.G.: Exploiting model uncertainty estimates for safe dynamic control learning. In: Advances in Neural Information Processing Systems 9 (NIPS), pp. 1047–1053 (1996)

Schwartz, A.: A reinforcement learning method for maximizing undiscounted rewards. In: International Conference on Machine Learning, ICML (1993)

Sehnke, F., Osendorfer, C., Rückstieß, T., Graves, A., Peters, J., Schmidhuber, J.: Parameter-exploring policy gradients. Neural Networks 21(4), 551–559 (2010)

Senoo, T., Namiki, A., Ishikawa, M.: Ball control in high-speed batting motion using hybrid trajectory generator. In: Proceedings of the IEEE International Conference on Robotics and Automation (ICRA), pp. 1762–1767 (2006)

Sentis, L., Khatib, O.: Synthesis of whole-body behaviors through hierarchical control of behavioral primitives. International Journal of Humanoid Robotics 2(4), 505–518 (2005)

Shone, T., Krudysz, G., Brown, K.: Dynamic manipulation of kendama. Tech. rep., Rensselaer Polytechnic Institute (2000)

Silver, D., Bagnell, J.A., Stentz, A.: High performance outdoor navigation from overhead data using imitation learning. In: Proceedings of Robotics: Science and Systems Conference, R:SS (2008)

Silver, D., Bagnell, J.A., Stentz, A.: Learning from demonstration for autonomous navigation in complex unstructured terrain. International Journal of Robotics Research 29(12), 1565–1592 (2010)

Simon, H.A.: The Shape of Automation for Men and Management. Harper & Row, New York (1965)

Smart, W.D., Kaelbling, L.P.: A framework for reinforcement learning on real robots. In: Proceedings of the National Conference on Artificial Intelligence/Innovative Applications of Artificial Intelligence, AAAI/IAAI (1998)

Smart, W.D., Kaelbling, L.P.: Effective reinforcement learning for mobile robots. In: Proceedings of the IEEE International Conference on Robotics and Automation, ICRA (2002)

Soni, V., Singh, S.: Reinforcement learning of hierarchical skills on the sony aibo robot. In: Proceedings of the International Conference on Development and Learning, ICDL (2006)

Sorg, J., Singh, S.P., Lewis, R.L.: Reward design via online gradient ascent. In: Advances in Neural Information Processing Systems 23 (NIPS), pp. 2190–2198 (2010)

Spall, J.C.: Introduction to Stochastic Search and Optimization, 1st edn. John Wiley & Sons, Inc., New York (2003)

Strens, M., Moore, A.: Direct policy search using paired statistical tests. In: Proceedings of the 18th International Conference on Machine Learning, ICML (2001)

Stulp, F., Sigaud, O.: Path integral policy improvement with covariance matrix adaptation. In: Proceedings of the International Conference on Machine Learning, ICML (2012)

Stulp, F., Theodorou, E., Kalakrishnan, M., Pastor, P., Righetti, L., Schaal, S.: Learning motion primitive goals for robust manipulation. In: Proceedings of the IEEE/RSJ International Conference on Intelligent Robots and Systems (IROS), pp. 325–331 (2011)

Sumners, C.: Toys in Space: Exploring Science with the Astronauts. McGraw-Hill (1997)

Sutton, R.S.: Integrated architectures for learning, planning, and reacting based on approximating dynamic programming. In: Proceedings of the International Machine Learning Conference (ICML), pp. 9–44 (1990)

Sutton, R.S., Barto, A.G.: Reinforcement Learning. MIT Press, Boston (1998)

Sutton, R.S., Barto, A.G., Williams, R.J.: Reinforcement learning is direct adaptive optimal control. In: American Control Conference, pp. 2143–2146 (1991)

Sutton, R.S., McAllester, D., Singh, S., Mansour, Y.: Policy gradient methods for reinforcement learning with function approximation. In: Advances in Neural Information Processing Systems 12 (NIPS), pp. 1057–1063 (1999)

Sutton, R.S., Koop, A., Silver, D.: On the role of tracking in stationary environments. In: Proceedings of the International Conference on Machine Learning, ICML (2007)

Svinin, M.M., Yamada, K., Ueda, K.: Emergent synthesis of motion patterns for locomotion robots. Artificial Intelligence in Engineering 15(4), 353–363 (2001)

Tadepalli, P., Ok, D.: H-learning: A reinforcement learning method to optimize undiscounted average reward. Tech. Rep. 94-30-1, Department of Computer Science, Oregon State University (1994)

Takenaka, K.: Dynamical control of manipulator with vision: "cup and ball" game demonstrated by robot. Transactions of the Japan Society of Mechanical Engineers C 50(458), 2046–2053 (1984)

Tamei, T., Shibata, T.: Policy gradient learning of cooperative interaction with a robot using user's biological signals. In: Proceedings of the International Conference on Neural Information Processing, ICONIP (2009)

Tamošiūnaitė, M., Nemec, B., Ude, A., Wörgötter, F.: Learning to pour with a robot arm combining goal and shape learning for dynamic movement primitives. Robotics and Autonomous Systems 59, 910–922 (2011)

Taylor, M.E., Whiteson, S., Stone, P.: Transfer via inter-task mappings in policy search reinforcement learning. In: Proceedings of the 6th International Joint Conference on Autonomous Agents and Multiagent Systems, AAMAS (2007)

Tedrake, R.: Stochastic policy gradient reinforcement learning on a simple 3D biped. In: Proceedings of the International Conference on Intelligent Robots and Systems, IROS (2004)

Tedrake, R., Zhang, T.W., Seung, H.S.: Stochastic policy gradient reinforcement learning on a simple 3d biped. In: Proceedings of the IEEE International Conference on Intelligent Robots and Systems (IROS), pp. 2849–2854 (2004)

Tedrake, R., Zhang, T.W., Seung, H.S.: Learning to walk in 20 minutes. In: Proceedings of the Yale Workshop on Adaptive and Learning Systems (2005)

Tedrake, R., Manchester, I.R., Tobenkin, M.M., Roberts, J.W.: LQR-trees: Feedback motion planning via sums of squares verification. International Journal of Robotics Research 29, 1038–1052 (2010)

Tesauro, G.: TD-gammon, a self-teaching backgammon program, achieves master-level play. Neural Computation 6(2), 215–219 (1994)

Tesch, M., Schneider, J.G., Choset, H.: Using response surfaces and expected improvement to optimize snake robot gait parameters. In: Proceedings of the IEEE/RSJ International Conference on Intelligent Robots and Systems (IROS), pp. 1069–1074 (2011)

Theodorou, E.A., Buchli, J., Schaal, S.: Reinforcement learning of motor skills in high dimensions: A path integral approach. In: Proceedings of the IEEE International Conference on Robotics and Automation (ICRA), pp. 2397–2403 (2010)

Thrun, S.: An approach to learning mobile robot navigation. Robotics and Autonomous Systems 15, 301–319 (1995)

Tokic, M., Ertel, W., Fessler, J.: The crawler, a class room demonstrator for reinforcement learning. In: Proceedings of the International Florida Artificial Intelligence Research Society Conference, FLAIRS (2009)

Toussaint, M., Goerick, C.: Probabilistic inference for structured planning in robotics. In: Proceedings of the IEEE/RSJ International Conference on Intelligent Robots and Systems, IROS (2007)

Toussaint, M., Storkey, A., Harmeling, S.: Expectation-Maximization methods for solving (PO)MDPs and optimal control problems. In: Inference and Learning in Dynamic Models. Cambridge University Press (2010)

Touzet, C.: Neural reinforcement learning for behaviour synthesis. Robotics and Autonomous Systems, Special Issue on Learning Robot: The New Wave 22(3-4), 251–281 (1997)

Towell, C., Howard, M., Vijayakumar, S.: Learning nullspace policies. In: Proceedings of the IEEE International Conference on Intelligent Robots and Systems, IROS (2010)

Tsitsiklis, J.N., Van Roy, B.: An analysis of temporal-difference learning with function approximation. IEEE Transactions on Automatic Control 42(5), 674–690 (1997)

Tyldesley, D., Whiting, H.: Operational timing. Journal of Human Movement Studies 1, 172–177 (1975)

Uchibe, E., Asada, M., Hosoda, K.: Cooperative behavior acquisition in multi mobile robots environment by reinforcement learning based on state vector estimation. In: Proceedings of the IEEE International Conference on Robotics and Automation, ICRA (1998)

Ude, A., Gams, A., Asfour, T., Morimoto, J.: Task-specific generalization of discrete and periodic dynamic movement primitives. IEEE Transactions on Robotics 26(5), 800–815 (2010)

Urbanek, H., Albu-Schäffer, A., van der Smagt, P.: Learning from demonstration repetitive movements for autonomous service robotics. In: Proceedings of the IEEE/RSJ International Conference on Intelligent Robots and Systems (IROS), vol. 4, pp. 3495–3500 (2004)

Vlassis, N., Toussaint, M., Kontes, G., Piperidis, S.: Learning model-free robot control by a Monte Carlo EM algorithm. Autonomous Robots 27(2), 123–130 (2009)

Wang, B., Li, J.W., Liu, H.: A heuristic reinforcement learning for robot approaching objects. In: Proceedings of the IEEE Conference on Robotics, Automation and Mechatronics (2006)

Welling, M.: The Kalman filter. Lecture Notes (2010)

Whitman, E.C., Atkeson, C.G.: Control of instantaneously coupled systems applied to humanoid walking. In: IEEE-RAS International Conference on Humanoid Robots, HUMANOIDS (2010)

Wikipedia, Fosbury flop (2013), http://en.wikipedia.org/wiki/FosburyFlop, http://en.wikipedia.org/wiki/FosburyFlop

Willgoss, R.A., Iqbal, J.: Reinforcement learning of behaviors in mobile robots using noisy infrared sensing. In: Proceedings of the Australian Conference on Robotics and Automation (1999)

Williams, R.J.: Simple statistical gradient-following algorithms for connectionist reinforcement learning. Machine Learning 8, 229–256 (1992)

Wulf, G.: Attention and motor skill learning. Human Kinetics, Champaign, IL (2007)

Yamaguchi, J., Takanishi, A.: Development of a biped walking robot having antagonistic driven joints using nonlinear spring mechanism. In: Proceedings of the IEEE International Conference on Robotics and Automation (ICRA), pp. 185–192 (1997)

Yasuda, T., Ohkura, K.: A reinforcement learning technique with an adaptive action generator for a multi-robot system. In: Proceedings of the International Conference on Simulation of Adaptive Behavior, SAB (2008)

Yoshikawa, T.: Manipulability of robotic mechanisms. The International Journal of Robotics Research 4(2), 3–9 (1985)

Youssef, S.M.: Neuro-based learning of mobile robots with evolutionary path planning. In: Proceedings of the ICGST International Conference on Automation, Robotics and Autonomous Systems (ARAS) (2005)

Zhou, K., Doyle, J.C.: Essentials of Robust Control. Prentice Hall (1997)

Ziebart, B.D., Maas, A., Bagnell, J.A., Dey, A.K.: Maximum entropy inverse reinforcement learning. In: Proceedings of the AAAI Conference on Artificial Intelligence (AAAI), pp. 1433–1438 (2008)

Zucker, M., Bagnell, J.A.: Reinforcement planning: RL for optimal planners. In: Proceedings of the IEEE International Conference on Robotics and Automation, ICRA (2012)

Index